T0291129

Environmental Law for Sustainable Construction

Environmental Law for Sustainable Construction

A guide for construction, engineering and architecture professionals

Francine Baker and Jennifer Charlson

Published by Emerald Publishing Limited, Floor 5, Northspring, 21-23 Wellington Street, Leeds LS1 4DL.

Other ICE Publishing titles:
Environmental Impact Assessment Handbook, Third edition
Barbara Carroll, Josh Fothergill, Jo Murphy and Trevor Turpin.
ISBN 978-0-7277-6141-5
Building Regulations, Codes and Standards: A guide for safe, sustainable and healthy development
Mark Key. ISBN 978-0-7277-6535-2
Planning Resilient Infrastructure Systems
Alexander Hay. ISBN 978-0-7277-6562-8

www.icebookshop.com
A catalogue record for this book is available from the British Library

ISBN 978-0-7277-6645-8

Cover photo: An aerial view of the buildings and dormitories of the University of York's Campus East. Clare Jackson/Alamy Stock Photo

Commissioning Editor: Michael Fenton
Development Editor: Cathy Sellars
Production Editor: Sirli Manitski

Typeset by Manila Typesetting Company
Index created by Madelon Nanninga-Fransen
Printed and bound in Great Britain by Bell and Bain, Glasgow

Contents

About the authors

Dr Francine Baker PhD, MA (1st cl. Hons), LLB, FHEA, FCICES, MCIArb

Dr Baker was admitted as a barrister and solicitor of the Supreme Court of Victoria, Australia, in 1991, and then to the High Court of Australia. She has worked for government authorities and private firms in Australia, and as a solicitor of the Supreme Court of England and Wales since 2004. Her areas of professional legal and teaching experience include contentious and non-contentious property, planning, environmental and construction law.

Since 2008, Dr Baker has also taught the above areas of law for construction and engineering courses at both undergraduate degree and postgraduate levels at various universities, including as a senior lecturer at London South Bank, where she was also director of their undergraduate quantity surveying and commercial management course, for the University of Reading and Oxford Brookes University; and currently for the University of East London. She has also been an external examiner of other universities' built environment courses, monitored overseas university franchises, and taught and directed an international CIArb and CICES accredited postgraduate course in construction law and dispute resolution. She has an active research portfolio and was a member of an international scientific advisory committee on waste management and environmental impact 2020 to 2022. Dr Baker's experience has also included membership of local government research ethics and university committees.

She is a member of Institution of Civil Engineers editorial panel for the *Management, Procurement and Law* journal, a member of the Environmental Law Foundation and of the United Kingdom Environmental Lawyers Association, and a patron of the Oxford Botanic Gardens.

Dr Jennifer Charlson PhD, MA (Oxon), MBA, CEng, FIET

Dr Charlson qualified as a solicitor of the Supreme Court of England and Wales in 1998. She has advised on contentious and non-contentious construction law in both a solicitor's practice and in-house for a leading UK support services and construction company. Her MBA is in legal practice.

As a senior university lecturer in a school of architecture and the built environment for over 10 years, she lectured on courses including an MSc course in construction law and dispute resolution. Her research specialism is construction law, encompassing the disciplines of the UK legal framework, environmental and planning law, procurement and dispute resolution. Her PhD is in construction law.

Dr Charlson previously worked for the BBC as a projects engineer. She is a graduate of the University of Oxford in engineering science, qualified as a chartered engineer and is a Fellow of the Institution of Engineering and Technology. She is an editorial panel member of the Institution of Civil Engineers' *Management, Procurement and Law* journal and an external examiner of an LLM construction law and practice course.

Glossary

AQS	air quality strategy
BAT	best available techniques
BIM	building information modelling
BNG	biodiversity net gain
Brexit	the withdrawal of the United Kingdom from the European Union
CA	Court of Appeal
CAR	contractors' all risks
CBD	carrier, broker and dealer
CCC	Climate Change Committee
CDE	construction, demolition and excavation
CEAP	circular economy action plan
CEP	circular economy package
CIL	Community Infrastructure Levy
CJEU	Court of Justice of the European Union
CLU	certificate of lawful use
COP	Conference of the Parties
CHSR	The Conservation of Habitats and Species (Amendment) Regulations 2017
COSHH	Control of Substances Hazardous to Health
CPA	Control of Pollution (Amendment) Act 1989
C&I	commercial and industrial
DAS	design and access statement
Defra	Department for Environment, Food and Rural Affairs
DLUHC	Department for Levelling Up, Housing and Communities
DMPO	The Town and Country Planning (Development Management Procedure) (England) Order 2015
DVSA	Driver and Vehicle Standards Agency
D&O	directors' and officers'

EA	Environment Agency
EC	European Community
EEC	European Economic Community
EIA	environmental impact assessment
EOR	environmental outcomes report
EN	enforcement notice
EP	environmental permit
EPA	Environmental Protection Act 1990
EPR	The Environmental Permitting (England and Wales) Regulations 2016
ES	environmental statement
EU	European Union
EVP	engineered into the void permanently
FDC	flood defence consent
FTT	First Tier Tribunal
GHG	greenhouse gas
GPDO	The Town and Country Planning (General Permitted Development) (England) Order 2015
HM	His Majesty's
HMRC	HM Revenue & Customs
HMO	house in multiple occupation
HSR	The Conservation of Habitats and Species Regulations 2017
IBC	intermediate bulk container
IDB	Internal Drainage Board
IPEIA	The Infrastructure Planning (Environmental Impact Assessment) Regulations 2017
JCT	Joint Contracts Tribunal
JR	judicial review
LDC	lawful development certificate
LPA	local planning authority
MoU	memorandum of understanding

NE	Natural England
NEC	New Engineering Contract (ICE)
NMP	noise management plan
NSIP	nationally significant infrastructure project
NPPF	National Planning Policy Framework
NRBW	Natural Resources Body for Wales
NRW	Natural Resources Wales
NWFD	non-waste framework directive
OECD	Organisation for Economic Co-operation and Development
OEP	Office for Environmental Protection
PD rights	permitted development rights
PPE	personal protective equipment
PI	professional indemnity
PPG	Planning Policy Guidance
RA	Reservoirs Act 1975
RAMSAR	Convention on Wetlands (named after the city of Ramsar in Iran, where the convention was signed in 1971)
REACH	registration, evaluation, authorisation and restriction of chemicals
REWDT	resource efficiency and waste reduction targets
RPS	regulatory position statement
SACs	special areas of conservation
SDGs	Sustainable Development Goals
SoS	Secretary of State
SPAs	special protection areas
SSSI	Site of Special Scientific Interest
SuDS	sustainable drainage systems
sui generis	in a class of its own
TCA	EU–UK Trade and Cooperation Agreement
TCLP	The Chancery Lane Project

TCPA	Town and Country Planning Act 1990
TCPEIA	The Town and Country Planning (Environmental Impact Assessment) Regulations 2017
WCF	The Water Supply (Water Fittings) Regulations 1999
UK	United Kingdom of Great Britain and Northern Ireland
UN	United Nations
UNFCCC	UN Framework Convention on Climate Change
WAC	waste acceptance criteria
WEEE	waste electrical and electronic equipment
WPA	waste planning authority
WRA	Water Resources Act 1991
WRAS	Water Regulations Advisory Scheme
WR	The Waste (England and Wales) Regulations 2011
WCA	Wildlife and Countryside Act 1981

Baker F and Charlson J
ISBN 978-0-7277-6645-8
https://doi.org/10.1680/elsc.66458.001

Introduction

Francine Baker

Given the international awareness of the impact of construction and other human activities on the environment and its biodiversity, the scope of environmental law has much expanded in recent years. The increasing number of severe weather-related incidents also requires companies within the construction sector to include climate risk management strategies for each of their projects and as part of their corporate framework. This requires knowledge of various areas of law, and increasingly of environmental law.

This book chiefly concerns environmental law in England. It also refers to law which concerns the UK, as well as to law which applies to both England and Wales. However, readers based in Scotland, Wales and Northern Ireland should check the applicable law in their region for differences.

It is intended to be a working guide for construction professionals, and so it is not written for lawyers. The term ‘construction’ is used in this book in a broad sense to encompass all the stages involved in the production of a structure, as well as all relevant activities and professions involved. It therefore includes, but is not limited to, the involvement of developers, contractors, architects, planners, civil engineers, mechanical and electrical engineers, site engineers, quantity and land surveyors, commercial managers, project managers and site managers, construction managers and cost estimators.

This book also considers the main interactions between planning law and environmental law. The areas of both environmental law and planning law have expanded considerably in the UK in the past 20 years, partly in recognition of climate change and the importance of providing a sustainable future for our world as reflected by the United Nations Sustainable Development Goals (discussed below). The result is that environmental issues and sustainability goals have taken a front seat in the day-to-day operation and management of construction projects. Problems on construction sites may develop quickly, and it is often unclear who is responsible for environmental law issues. Therefore, this book will be particularly relevant to all construction professionals within the industry. However, since Brexit, new relevant law is on the horizon, some of which is referred to here, but has yet to be enacted or drafted. This book may, therefore, be updated with appendixes.

The objectives of environmental law

Environmental law is an evolving area of law which developed following the aftermath of the UK’s industrial revolution. The coal-powered industrial revolution of the mid-18th century to the 19th century had a destructive impact on air, water, vegetation, fish stocks and rivers in England (Christman, 2013). Various laws were passed in the 19th century to address discrete aspects of a polluted environment, such as the Smoke Nuisance Abatement Act 1853, which sought to protect the quality of air, the Alkali Act 1863, which required chemical emissions to be stopped or diluted to protect vegetation, and the Rivers Pollution Act 1876, which aimed to address stream and river pollution from contamination by sewerage and other deposits. By the 20th century, there was finally a Clean Air Act 1956, and the first waste management legislation – the Deposit of Poisonous Waste Act 1972, (the same year that the UK joined the European Union) and the Control of Pollution Act 1974. However, modern environmental law may be said to have started with the Environmental Protection Act 1990, which superseded the latter legislation. This Act was needed to effectively regulate the collection, delivery and disposal of waste, and to address a range of causes of pollution and nuisances affecting our environment, whether on land or in the sea.

Modern environmental law in the UK includes both common law and legislation. Common law (case law) is based on the reasons given by judges for their decisions on cases they have heard in court. While such decisions are based on individual cases, they can be applied as precedent in future cases. The other main UK legislation is law that has been passed by the UK Parliament. Primary legislation is the main law passed by Parliament and is called 'acts' (also known as statutes). Secondary legislation is law made by persons or organisations given authority to draft law by a particular act of Parliament. Such legislation is called a 'statutory instrument', and includes orders, codes, regulations and rules. Chapter 1 explores the legal framework of environmental law in more detail.

Environmental law seeks to protect the environment, which is understood as the natural world we live in. It therefore seeks to protect the quality of the air we breathe, and naturally formed bodies of water, such as rivers, waterfalls, the seas, and lakes and streams. It seeks to protect the biodiversity (the variety and variability of plant and animal life) of the environment. It can include protozoa, beetle larvae, ants, earthworms, arachnids, rodents and more, as well as naturally formed minerals and mountains, and all kinds of naturally occurring vegetation, including fungi, bacteria, forests and marshes, and all kinds of species of creatures. The need for its protection extends to include the habitats of all wildlife such as animals, reptiles, birds and insects, including bees, spiders, ticks, centipedes and marine life. The inhabitants of the natural world are part of ecosystems that produce our food, the air we breathe and the water we drink. Environmental law seeks to protect the planet, whereas the science of ecology provides knowledge about the interdependence between people and nature that is vital for food production, maintaining clean air and water, and sustaining biodiversity.

Through damaging the environment, we contribute to destroying a viable home for us on this planet. The term 'biodiversity' is mentioned throughout in this book, but in particular see the law discussed in Section 2.12.5.3 in Chapter 2 and Sections 3.6.1–3.6.4 in Chapter 3.

The impact of the construction industry on the environment

The environmental impact of both onsite and offsite construction has been recognised in planning law as well as environmental law, so that the relevant local planning authority or Secretary of State, as it may be, may not grant planning permission, or a permit to allow certain work, if the development or activity is likely to damage a protected area or site, species or other wildlife, or is likely to damage the onsite or surrounding environment. This is discussed in Chapter 2, Section 2.12.4, and the following sections.

The noise and light that construction sites generate may not directly affect any animals present, but it can impact their breeding and feeding behaviours, which may affect the future numbers of the species (see Chapter 3, Section 3.3.7.1; Chapter 6, Section 6.11; and Chapter 7, Section 7.6). The erection or demolition of buildings could cause the separation of species and their habitats, impacting on the dynamics of the ecosystem. These factors can contribute to a decline in population species and biodiversity.

Risk assessments and environmental impact assessments may be legally required for planning permission purposes where the area for development or nearby area concerns natural habitats and the work involves use of natural resources, in particular land, soil, water and biodiversity in the area and its underground (see Chapter 3, particularly Section 3.3.3.3). More recently, the Environment Act 2021 requires that there is at least one 'long-term' target set for each priority area, that is, to improve air and water quality, reduce waste, improve biodiversity, and improve resource efficiency and waste reduction. It also requires targets to be set to cut exposure to fine particulate matter ($PM_{2.5}$) and improve species abundance, and requires biodiversity net gain requirements, which will need to be met before planning permission for developments is agreed, that is, once the provisions become operational in late 2023 or 2024. This is discussed in Chapter 3, Section 3.6.1. The UK government published its Environmental Improvement Plan in January 2023 setting out how it proposes to achieve these targets (CIEEM, 2022).

The Environmental Protection Act 1990 (as amended) places an obligation on local authorities to identify contaminated land. Land is considered contaminated if it has been affected by a pollutant. Contaminated land is often found on sites that have been used for industrial purposes, as well as those known as 'brownfield sites'. Chapter 4 explains the contaminated land regime and the legal obligations of construction industry professionals when dealing with contaminated land. Construction is a major contributor to air pollution levels. According to the Centre for Low Emission Construction (2019, p. 3),

> The 2019 London Atmospheric Emissions Inventory (LAEI) shows that approximately 30% of particulate matter (PM_{10}) emissions come from construction, along with 8% of fine particulate matter ($PM_{2.5}$) and 4% of nitrogen oxides.

The law relating to air pollution is discussed in Chapter 7.

Impact of Brexit on environmental law

Under the European Union (Withdrawal) Act 2018, most European Union (EU) law, including environmental law and laws that affect business, worker's rights, food standards, industry and animal rights, was kept (retained) for the UK following the end of the post-Brexit transition period, for the practical reason that to do otherwise would have been unmanageable at the time. Under section 6(3) of the Act, the UK courts are required, when interpreting unmodified retained EU law, to follow retained domestic case law, retained EU case law and retained general principles of EU law until a relevant UK court departs from it or it is modified by legislation.

However, the government introduced a proposal for legislation called the Retained EU Law (Revocation and Reform) Bill into Parliament on 22 September 2022. The purpose of the bill is to provide provisions which would enable the government to mend EU retained law or to remove it from the UK legal system. The bill aims to review and repeal or assimilate EU retained law by the end of 2023. However, the government could delay the deadline until 2026, as the process of reviewing over 2400 documents is a huge undertaking. What law will be kept and what will be removed (sunsetted) is unclear.

The impact of any changes to UK law, including environment law, on those affected have not yet been assessed by the government. However, the independent Better Regulation watchdog considers that it is reasonable to expect the government to properly consider the impacts of such changes (Regulatory Policy Committee, 2022). Yet the current bill does not provide for the creation of secondary legislation such as regulations to address impacts. However, the bill is still proceeding through the Houses of Parliament, and it is possible that changes may be made before its final form. Brexit and UK–EU trade relations are discussed in Chapter 1, Sections 1.2.2–1.2.4. As EU directives may still be relevant for the purposes of understanding and interpreting certain provisions of retained EU-derived domestic legislation, there are further references to Brexit regarding the impact on waste management in Chapter 5, Sections 5.2.4 and 5.9.1, and to exit regulations regarding floods and water legislation in Chapter 6, Section 6.2.1.

Sustainable Development Goals (SDGs)

The term 'sustainability' is commonly used in the construction industry to reflect courses of action, whether technical or performance related, to support a sustainable environment. The Brundtland Report (United Nations, 1987) defines sustainability as a balance between the economic, social and environmental elements of society in a manner that meets the needs of the present without compromising the ability of future generations to meet their needs; this definition is generally accepted.

In New York, in September 2015, the 193 countries that comprise the United Nations Assembly agreed a plan involving 17 interdependent Sustainable Development Goals (SDGs) and 169 targets to be reached by 2030. These goals linked social, economic and environmental aspects of the Millennium Development Goals agreed at the Millennium Summit of the United Nations in 2000, in recognition that action in one area affects outcomes in other areas (United Nations, 2022a). In 2015, the SDGs were adopted and ratified (signed) by 187 parties in an agreement known as the Paris Accord at the 21st Conference of the Parties to the United Nations Framework Convention on Climate Change, also known as COP21 (see Chapter 1, Section 1.2.3.2). The SDGs are set out below.

- Goal 1: End poverty in all its forms everywhere.
- Goal 2: End hunger, achieve food security and improved nutrition, and promote sustainable agriculture.
- Goal 3: Ensure healthy lives and promote wellbeing for all at all ages.
- Goal 4: Ensure inclusive and quality education for all and promote lifelong learning.
- Goal 5: Achieve gender equality and empower all women and girls.
- Goal 6: Ensure availability and sustainable management of water and sanitation for all.
- Goal 7: Ensure access to affordable, reliable, sustainable and modern energy for all.
- Goal 8: Promote sustained, inclusive and sustainable economic growth, full and productive employment and decent work for all.

- Goal 9: Build resilient infrastructure, promote inclusive and sustainable industrialisation, and foster innovation.
- Goal 10: Reduce inequality within and among countries.
- Goal 11: Make cities and human settlements inclusive, safe, resilient and sustainable.
- Goal 12: Ensure sustainable consumption and production patterns.
- Goal 13: Take urgent action to combat climate change and its impacts.
- Goal 14: Conserve and sustainably use the oceans, seas and marine resources for sustainable development.
- Goal 15: Protect, restore and promote sustainable use of terrestrial ecosystems, sustainably manage forests, combat desertification, halt and reverse land degradation, and halt biodiversity loss.
- Goal 16: Promote peaceful and inclusive societies for sustainable development, provide access to justice for all, and build effective, accountable and inclusive institutions at all levels.
- Goal 17: Strengthen the means of implementation and revitalise the global partnership for sustainable development.

All the SDGs are important and function interdependently to address climate change in every industry sector. They also inform the building of an appropriate infrastructure for a sustainable future (Thacker *et al.*, 2019).

The UK construction industry aligns its activities with the goal of supporting the SDGs through legal compliance with various standards and codes of practice, and through advisory and certification organisations (Archdesk, 2022; BRE, 2023; 17 Global Goals, 2020). In July 2022, the largest privately owned construction company in the UK, Laing O'Rourke, signed up as one of the 17 founding members of the new ConcreteZero campaign, an international non-profit initiative led by the Climate Group in partnership with the World Green Building Council and World Business Council for Sustainable Development (Laing O'Rourke, 2022). Its pledge is to achieve 100% net-zero concrete by 2050, and the use of 30% low-emission concrete by 2025 and 50% by 2030. The relationship between open and fair competition and sustainable development is discussed in Chapter 1, Section 1.2.3.4.

The construction industry's professional institutions have also played a leading role in promoting the implementation of SDGs to its members and to the public on their websites. The following provide a few examples. The Institution of Civil Engineers (ICE) launched a Sustainability Road Map in 2019 to engage with the SDGs and report and address the SDG impact of their infrastructure projects and programmes (ICE, 2019). The Royal Institution of Chartered Surveyors (RICS, 2022) stated that it has been working with the United Nations on critical issues facing the construction industry to assist in the implementation of the SDGs. In September 2022, the Chartered Institution of Civil Engineering Surveyors (CICES) began a consultation on a white paper with the aim of how best to implement its Sustainability Strategy 2021–2030 (CICES, 2022). The Chartered Institute of Building (CIOB) stated in December 2021 that 'The UN social sustainability goals on social sustainability fit closely with our own vision, mission and values to improve the quality of life for those who use and create our built environment.' (CIOB, 2021).

UK net-zero target and procurement goals

Prior to 2019, the UK was committed to reducing its net greenhouse gas emissions by at least 80% of their 1990 levels. However, in 2019, Parliament amended the Climate Change Act 2008 to require that the UK's net emissions of greenhouse gases be reduced by 100% relative to 1990 levels by 2050.

> Net zero means any emissions would be balanced by schemes to offset an equivalent amount of greenhouse gases from the atmosphere, such as planting trees or using technology like carbon capture and storage. (BEIS, 2019)

> Carbon dioxide (CO_2) emissions in the UK are provisionally estimated to have increased by 6.3% in 2021 from 2020, to 341.5 million tonnes (Mt), and total greenhouse gas emissions by 4.7% to 424.5 million tonnes carbon dioxide equivalent ($MtCO_2e$). Compared to 2019, the most recent pre-pandemic year, 2021 CO_2 emissions are down 5.0% and total greenhouse gas emissions are down 5.2%. Total greenhouse gas emissions were 47.3% lower than they were in 1990. (BEIS, 2021)

The unit $MtCO_2e$ is a million metric tonnes of carbon dioxide equivalent (CO_2e). The carbon dioxide equivalent is a measure used to compare the global warming potential (GWP) of different greenhouse gases (GHG) by converting the amount of a GHG to the equivalent amount of CO_2 that would have the same atmospheric impact.

On 20 April 2021, following advice from the UK Climate Change Committee's sixth carbon budget on 9 December 2020, the UK government announced that it 'will set the world's most ambitious climate change target' to reduce emissions by 78% by 2035 compared to 1990 levels (Dray, 2021). The UK's Net Zero Strategy was launched in October 2021 (DESNZ and BEIS, 2021). It sets out policies and proposals for decarbonising all sectors of the UK economy to meet the net-zero target by 2050.

However, the strategy was being reviewed following a mandatory order given by the High Court in *Friends of the Earth Ltd & Ors, R (On the Application Of) v Secretary of State for Business, Energy and Industrial Strategy* [2022] EWHC 1841 for the government to lay a fresh report before Parliament by March 2023 to fully address its legal obligations under section 14 of the Climate Change Act 2008 as to how it would achieve its emissions targets. An independent review of the strategy was commissioned, with a call for evidence. The review of the Strategy was published on 13 January 2023 (Skidmore, 2023), and it includes recommendations on how the government can move faster to implement the delivery of its net-zero goals. A revised net-zero strategy was published on 30 March 2023 (ESNZ, 2023).

The UK government's National Procurement Policy Statement (HMG, 2021) sets out the strategic priorities for public procurement, and how contracting authorities can support the delivery of these through tackling climate change and reducing waste by

- contributing to the UK government's legally binding target to reduce greenhouse gas emissions to net zero by 2050

- reducing waste, improving resource efficiency and contributing to the move towards a circular economy
- identifying and prioritising opportunities in sustainable procurement to deliver additional environmental benefits, for example enhanced biodiversity, through the delivery of the contract.

Environmental and planning law work together to support the implementation of these priorities. Environmental law is a relevant tool for enabling the move towards a circular economy and the implementation of sustainable procurement goals. Planning law plays a part in the creation and maintenance of a framework within which sustainable environments are encouraged and environmental law is enforced, as discussed in Chapters 2, 3 and 7.

Recent developments – environmental law

On 28 July 2022, the United Nations General Assembly (UNGA) adopted a resolution declaring that everyone on the planet has a right to a healthy environment (United Nations, 2022b). The UK Human Rights Act 1998 (HRA) which incorporated the European Convention of Human Rights does not refer to a human right to a healthy environment, although Article 8, 'the right to respect for ... privacy and family life', has been argued more recently in relation to air pollution cases (UK Parliament, 2022). However, although the discussion of human rights law is beyond the scope of this book, UNGA's recognition there is a relationship between human rights and the environment signals a new and developing area of law.

More developments in UK law include the Environment Act 2021. The Act requires more construction industry accountability regarding the preservation of biodiversity (discussed in Chapter 3), the reduction of water, air and land pollution, (referred to in Chapters 3 and 4), and the use and management of water and waste management (discussed in Chapters 6 and 5, respectively). The Act also relates to the management of contaminated or brownfield land (discussed in Chapters 4 and 7). Construction industry players have personal and managerial legal obligations to protect the environment from a condition or activity that interferes with the use or enjoyment of land or from its contamination.

A bill which may impact on the planning process and how various environmental impacts are assessed when it becomes law is the Levelling Up and Regeneration Bill. It was introduced to Parliament on 11 May 2022, and is expected to become an Act of Parliament during 2023. The bill's proposals, assuming they become law, will amend various legislation, including the Town and Country Planning Act 1990 and the Planning Act 2008, the Local Democracy, Economic Development and Construction Act 2009 and impact on environmental legislation (see part 6 section 152 of the bill). A feature of the bill is a reference to a consideration of the impact on or improving the economic, social and environmental wellbeing of some or all of the people in the relevant area, and so it aligns itself with the concept of sustainability.

Among the range of proposed changes to various legislation, the Levelling Up and Regeneration Bill proposes to simplify and speed up the local planning process (part 3 of the bill). However, part 6 of the bill only gives the Secretary of State the power to replace the current environmental impact assessment regimes with a simplified process involving an environmental outcomes report (EOR). For example, section 138 refers to EOR regulations yet to be made by the Secretary of State under this part, and states that the Secretary of State 'may' specify the EOR 'outcomes' relating to environmental protection. The use of a mandatory term such as 'should' or 'must' in this part would have required such outcomes to be produced; however, it has not been used here. The proposed EOR process, assuming the relevant regulations are made, will further prioritise protecting the environment when local authorities or the Secretary of State are considering development plans and planning applications. Part 6 section 149 also states that any EOR regulations made may interact 'with existing environmental assessment legislation or the Habitats Regulations'. The latter phrase is widely interpreted at section 152 of part 6. In addition, part 7, when law, will amend the Water Industry Act 1991 by requiring certain sewage disposal works to meet nutrient pollution standards set out in section 96F.

The bill also proposes to introduce a new infrastructure levy (part 4 of the bill) to be paid by developers when the property is sold, with the rates and thresholds likely to be set by local authorities, and the green regeneration of brownfield sites (discussed in Chapter 4) will be supported by changes to the compulsory purchase system. However, just how this bill, when it becomes law, will impact on the planning and decision-making process and environmental assessments and protections, as well as the financial impact on the construction industry, will become apparent in changes and regulations yet to make. Therefore, a clear picture of the impact of this bill on the industry is uncertain.

The proposals in the Levelling Up and Regeneration Bill, if it becomes law, and when certain proposals result in regulations, are likely to increase the extent of the current requirements set out in this book to avoid and mitigate any environmental impacts, as well as to further compensate for any damage to the environment when outcomes are not met.

Recent developments – implementing the law

The use of building information modelling (BIM) by the construction industry also contributes towards managing risks and protecting the environment. This is discussed in ICE's publications *BIM in Principle and in Practice*, and *BIM for Project Managers*, both by Peter Barnes. They explain that the process can increase productivity while reducing the use of resources at all stages of the project. The many building energy simulation tools available through BIM can ensure that the design efficiently optimises the use and reduces the waste of energy, water and other resources. Its ability to carry out a life-cycle analysis can provide a whole-life assessment of the environmental impact of products or services, which can then be used to amend elements of the design to reduce environmental impacts.

The increasing number of new rules and regulations being passed to address climate change and to minimise the impact of construction activities on the environment increases liability risk. Construction companies can ensure sufficient climate protection is already in place before work begins, and during all stages of the development. Any potential risks that could arise may be managed through appropriate insurance and the provision of environment-related insurance cover (discussed in Chapter 8).

Legal professionals are also creating new construction contract environment-friendly clauses which also assist with the achievement of the SDGs. For example, a range of climate clauses, case studies and a Net Zero Toolkit is provided by The Chancery Lane Project (2023) for the use of lawyers. Chapter 8, Section 8.8 briefly refers to insurance-focused climate change clauses provided by the project to assess risk.

On 26 July 2022, ICE's New Engineering Contract (NEC) was the first of the construction industry standard contracts to introduce a new secondary contract clause (called X29) to address climate change issues and to provide a climate change plan to assist the move to decarbonisation and help the industry achieve net-zero emissions and sustainability targets in the creation and operation of built assets (NEC, 2022).

This is timely, given the UK government's Guidance Note, *Promoting Net Zero Carbon and Sustainability in Construction* (Government Commercial Function, 2022), which consists of practical resources on decarbonisation for those procuring construction and infrastructure projects and programmes. This guidance applies to all central government departments, their executive agencies and non-departmental public bodies from the time of publication (September 2022) and contracting authorities within the wider public sector.

However, this book does not attempt to examine construction contract agreement clauses relevant to the implementation of SDGs. Neither does it deal with energy law, as this is a separate body of law from environmental law, and it is also a large and rapidly expanding area of law that deserves its own space. The same set of facts that concern environmental law may also involve breaches of health and safety law, which may be referred to in passing but is not a subject of this book. As this book does not cover construction law or health and safety law, it does not consider the Building Safety Act 2022. Rather, this book relates to key environmental law topics relevant to the mainstream construction industry, broadly defined, and the observance of which broadly contributes to the achievement of a sustainable environment.

The structure of the book

This book commences with a chapter providing an overview of the legal system in the UK in the context of Brexit. Chapter 2 then considers the legal requirements for the submission of planning applications for developments, their environmental aspects and the range of environmental permits needed for construction work to proceed. The main types of environmental impact-related assessments required for local authority or central government permission for work to proceed are considered in Chapter 3. The main legal requirements regarding contaminated land are dealt with in Chapter 4. The developing and expanding legislation concerning waste management is summarised in Chapter 5, and the regulation of water pollution is addressed in Chapter 6. The management of site hazards and nuisances and developments regarding air pollution is covered in Chapter 7. The last substantive chapter, Chapter 8, considers environmental-related insurance matters and potential issues for the industry.

Each chapter of the book commences with an introduction that provides a summary of the contents of the chapter. A list of academic literature and government website addresses as sources of further information is given at the end of each chapter. However, note that websites may be updated or changed from time to time. As this book has also been written to assist with the environmental aspects of project planning, there is a summary in the form of a chronological checklist in Chapter 9 (Proactive Project Planning).

While attempting to cover a range of environmental issues, the emphasis has been on producing a user-friendly summary guide with the aim to simplify some of the complexity and confusion that surrounds the area of environmental law. However, the authors make no claims to cover every aspect of environmental law, or for this book to be the 'final word' on the subject.

It is anticipated that this book will be used mainly as a reference manual which can be read selectively when information about a specific topic is required. For this reason, an index is provided to cross-reference the reader to relevant chapters.

The contents of this book are not to be interpreted as legal advice.

References

Parliamentary bills

Retained EU Law (Revocation and Reform) Bill. https://bills.parliament.uk/bills/3340 (accessed 09/03/2023).

Levelling Up and Regeneration Bill. https://bills.parliament.uk/bills/3155 (accessed 09/03/2023).

Statutes

Alkali Act 1863
Clean Air Act 1956
Climate Change Act 2008
Control of Pollution Act 1974
Deposit of Poisonous Waste Act 1972
Environment Act 2021
Environmental Protection Act 1990
European Union (Withdrawal) Act 2018
Human Rights Act 1998 (HRA)
Planning Act 2008
Rivers Pollution Act 1876
Smoke Nuisance Abatement Act 1853
The Local Democracy, Economic Development and Construction Act 2009
The Town and Country Planning Act 1990
Water Industry Act 1991

Case law

Friends of the Earth Ltd & Ors, R (On the Application Of) v Secretary of State for Business, Energy and Industrial Strategy [2022] EWHC 1841.

Journals

Christman B (2013) A brief history of environmental law in the UK. *Environmental Scientist* **22(4)**: Research Paper No. 2014-03. https://papers.ssrn.com/sol3/papers.cfm?abstract_id=2383906#:~:text=The%20first%20statutory%20response%20to%20industrial%20pollution%20was%20the%20Alkali%20Act%201863 (accessed 09/03/2023).

Thacker S, Adshead D and Fay M *et al.* (2019) Infrastructure for sustainable development. *Nature Sustainability* **2**: 324–331.

Books

Barnes P (2019) *BIM in Principle and in Practice*, 3rd edition. ICE Publishing, London, UK.

Barnes P (2020) *BIM for Project Managers: Digital Construction Management*, 3rd edition. ICE Publishing, London, UK.

Reports

Brundtland GH (1987) Our Common Future Report of the World Commission on Environment and Development, Geneva, UN-Dokument A/42/427. https://digitallibrary.un.org/record/139811?ln=en (accessed 11/05/23).

Websites

Archdesk (2022) Sustainable construction: how can industry support the UN Sustainable Development Goals? https://archdesk.com/blog/Sustainable-construction-how-can-industry-support-un-sustainable-develop (accessed 09/03/2023).

BEIS (Department for Business, Energy and Industrial Strategy) (2019) UK becomes first major economy to pass net zero emissions law. https://www.gov.uk/government/news/uk-becomes-first-major-economy-to-pass-net-zero-emissions-law (accessed 09/03/2023).

BEIS (2021) *2021 UK Greenhouse Gas Emissions*. Provisional figures, 31 March 2022, National Statistics. https://assets.publishing.service.gov.uk/government/uploads/system/uploads/attachment_data/file/1064923/2021-provisional-emissions-statistics-report.pdf (accessed 09/03/2023).

BRE (2023) Sustainable Development Goals. BREEAM. https://bregroup.com/products/breeam/sustainable-development-goals (accessed 09/03/2023).

CIEEM (Chartered Institute of Ecology and Environmental Management) (2022) UK government sets out environmental targets. https://cieem.net/uk-government-sets-out-environmental-targets (accessed 09/03/2023).

DESNZ and BEIS (Department for Energy Security and Net Zero and Department for Business, Energy and Industrial Strategy) (2021) *Net Zero Strategy: Build Back Greener*. Policy paper. https://www.gov.uk/government/publications/net-zero-strategy (accessed 09/03/2023).

Dray S (2021) Climate change targets: the road to net zero? Focus, 24 May. UK Parliament. https://lordslibrary.parliament.uk/climate-change-targets-the-road-to-net-zero (accessed 09/03/2023).

ESNZ (Department for Energy Security and Net Zero) (2023) *Powering Up Britain*. https://www.gov.uk/government/publications/powering-up-britain (accessed 26/04/23).

Evans A (2022) Sustainability white paper – your contribution. Chartered Institution of Civil Engineering Surveyors (CICES). https://www.cices.org/news/sustainability-a-call-to-action (accessed 09/03/2023).

Government Commercial Function (2022) *Promoting Net Zero Carbon and Sustainability in Construction*. Guidance Note. https://assets.publishing.service.gov.uk/government/uploads/system/uploads/attachment_data/file/1102389/20220901-Carbon-Net-Zero-Guidance-Note.pdf (accessed 09/03/2023).

HMG (His Majesty's Government) (2021) *National Procurement Policy Statement*. https://assets.publishing.service.gov.uk/government/uploads/system/uploads/attachment_data/file/990289/National_Procurement_Policy_Statement.pdf (accessed 09/03/2023).

ICE (2019) Engineering route map launched to deliver UN SDGs. https://www.ice.org.uk/news-and-insight/latest-ice-news/engineering-route-map-launched (accessed 09/03/2023).

Laing O'Rourke (2022) Liang O'Rourke signs up as founding member of ConcreteZero. https://www.laingorourke.com/company/press-releases/2022/laing-o-rourke-sign-up-as-founding-member-of-concretezero (accessed 09/03/2023).

NEC (2022) Introducing secondary option X29 climate change. https://www.neccontract.com/news/final-version-of-x29-released (accessed 09/03/2023).

Regulatory Policy Committee (2022) Retained EU Law (Revocation & Reform) Bill: RPC Opinion (Red-rated). https://www.gov.uk/government/publications/retained-eu-law-revocation-reform-bill-rpc-opinion-red-rated (accessed 09/03/2023).

RICS (Royal Institution of Chartered Surveyors) (2022) UN sustainable development. https://www.rics.org/about-rics/responsible-business/un-sustainable-development (accessed 09/03/2023).

Skidmore C (2023) *Mission Zero: Independent Review of Net Zero*. Department for Energy Security and Net Zero and Department for Business, Energy and Industrial Strategy. https://www.gov.uk/government/publications/review-of-net-zero (accessed 09/03/2023).

The Chancery Lane Project (2023) About The Chancery Lane Project. https://chancerylaneproject.org/about (accessed 09/03/2023).

Thorpe R (2021) Sustainability and the contribution we can make as an industry. Chartered Institute of Building (CIOB). https://www.ciob.org/blog/sustainability-contribution-industry (accessed 09/03/2023).

United Nations (1987) *Report of the World Commission on Environment and Development: Our Common Future*. Brundtland Report. https://sustainabledevelopment.un.org/content/documents/5987our-common-future.pdf (accessed 09/03/2023).

United Nations (2022a) The 17 goals. Department of Economic and Social Affairs, Sustainable Development. https://sdgs.un.org/goals (accessed 09/03/2023).

United Nations (2022b) UN General Assembly declares access to clean and healthy environment a universal human right. https://news.un.org/en/story/2022/07/1123482 (accessed 09/03/2023).

17 Global Goals (2020) Sustainable Development Goals, and the construction industry. https://17globalgoals.com/sustainable-development-goals-and-the-construction-industry (accessed 09/03/2023).

Baker F and Charlson J
ISBN 978-0-7277-6645-8
https://doi.org/10.1680/elsc.66458.013

Chapter 1
The UK environmental legal framework, regulators and advisers

Jennifer Charlson

1.1. Introduction

This chapter explains the UK environmental legal framework including European Union (EU) law, the EU–UK Trade and Cooperation Agreement (TCA), legislation and case law. The responsibilities of environmental regulators (local authorities, the Department for Environment, Food and Rural Affairs (Defra), the Environment Agency, Natural Resources Wales and the Office for Environmental Protection) are outlined, and the role of the environmental advisory bodies (Natural England and the Climate Change Committee) are explained.

1.1.1 UK environmental legal framework

The UK environmental legal framework includes EU law, the EU–UK TCA, and UK legislation and case law. The TCA contains a number of provisions relevant to environmental law. The European Union (Future Relationship) Act 2020 provides that UK law is amended as necessary to comply with the UK's obligations under the TCA.

Acts of Parliament are made by the UK Parliament (which comprises the House of Commons and the House of Lords), acting in its legislative role. The Climate Change Act 2008 contains a legally binding target committing the UK to net-zero greenhouse gas emissions by 2050. The Environment Act 2021 contains reforms aimed at protecting the environment now the UK has left the EU. Due to pressure on parliamentary time, Parliament may lay down a framework in an act (an 'enabling act') and then grant power to some other person or organisation to make the detailed regulations.

When a judge makes a decision, principles from previous cases will normally be applied. Under the English legal system, a judge is bound by earlier decisions (i.e. judicial binding precedent). The Technology and Construction Court handles disputes about buildings, engineering and surveying, including environmental claims.

1.1.2 Regulators and advisers

The sanction for breach of most environmental law is prosecution of an individual or company by the relevant regulator in the criminal courts. Penalties include fines and imprisonment.

Regulators can serve enforcement notices on operators requiring them to rectify breaches of environmental law. Failure to comply with enforcement notices can constitute a criminal offence. In some cases, regulators can shut down an operator's activities until the breach has been rectified.

1.2. European Union law

Sources of EU law include treaties, regulations, directives and decisions. The European Union (Withdrawal Agreement) Act 2020 paved the way for the UK to leave the EU, commonly known as 'Brexit', on 31 January 2020. The UK and EU agreed the EU–UK TCA on 24 December 2020. The TCA contains a number of provisions relevant to environmental law. It includes level-playing-field provisions to prevent distortions to trade and investment, for example, through state subsidies or reductions in environmental regulations standards. The European Union (Future Relationship) Act 2020 provides that UK law is amended as necessary to comply with the UK's obligations under the TCA.

1.2.1 Sources of EU law

The primary source of EU law is treaties. EU secondary legislation includes regulations, directives and decisions. Regulations are binding and are directly applicable to all EU member states. Directives are binding as to the result to be achieved but leave national authorities the choice of form and methods, although can be directly applicable in certain, narrow circumstances. Decisions are specific, binding and enforceable on member states.

1.2.2 UK leaving the EU (Brexit)

The European Communities Act 1972 facilitated the UK's entry into the European Economic Community (EEC) and recognised the legal validity and direct applicability of EEC law. The UK joined the EEC on 1 January 1973, and, in a 1975 referendum, 67% of people voted to stay in the EEC.

> **Box 1.1** *R v Secretary of State for Transport, ex parte Factortame Ltd (No. 2)* [1991] 1 AC 603
>
> In this case, the House of Lords acknowledged the supremacy of EEC law over the national law. Therefore, the doctrine of parliamentary sovereignty had been undermined, and UK courts were under a duty to override any rule of national law found to be in conflict with any directly enforceable rule of EEC law.

In the UK's EU membership referendum in June 2016, 51.9% of participants who voted (on the basis of a 72.2% turnout of more than 30 million people) were in favour of leaving the EU. The European Union (Withdrawal Agreement) Act 2020 paved the way for the UK to leave the EU, commonly known as Brexit, on 31 January 2020 (Charlson, 2021a). EU law in force on that date became part of the UK's domestic legal framework, pursuant to the European Union (Withdrawal) Act 2018, as a new category of retained EU law.

1.2.3 EU–UK Trade and Cooperation Agreement

The UK and EU agreed the TCA (HMG, 2020a) on 24 December 2020. The TCA, which was provisionally applied by the EU from 1 January 2021, was presented to Parliament in April 2021 and came into force on 1 May 2021. The TCA is structured in seven parts (HMG, 2020b).

- Part 1 covers the common and institutional provisions.
- Part 2 covers trade and other economic aspects of the relationship, such as aviation, energy, road transport and social security.
- Part 3 covers cooperation on law enforcement and criminal justice.
- Part 4 covers 'thematic' issues, notably health collaboration.
- Part 5 covers participation in EU programmes.
- Part 6 covers dispute settlement.
- Part 7 sets out final provisions.

1.2.3.1 TCA environmental law provisions

The TCA contains a number of provisions relevant to environmental law, including obligations on the parties

- not to reduce their level of environmental or climate protection below levels established by 31 December 2020
- to implement the UN Framework Convention on Climate Change (United Nations, 1992) and Paris Agreement (United Nations, 2015) and maintain effective carbon pricing systems
- to take measures to conserve biological diversity
- not to maintain anti-competitive subsidies except in certain circumstances, including for reasons of environmental protection
- not to allow subsidies for renewable energy that interfere with competitive markets and efficient interconnection.

1.2.3.2 UN Framework Convention on Climate Change and COP

The UN Framework Convention on Climate Change (UNFCCC) established an international environmental treaty to combat 'dangerous human interference with the climate system', in part by stabilising greenhouse gas concentrations in the atmosphere. The Paris Agreement is a legally binding international treaty on climate change adopted by 196 parties at COP21 in Paris. Its goal is to limit global warming to well below 2°C, preferably to 1.5°C, compared with pre-industrial levels.

The Conference of the Parties (COP) is the supreme decision-making body of the UNFCCC. A key task for the COP is to review the national communications and emission inventories submitted by the parties. Based on this information, the COP assesses the effects of the measures taken by the parties and the progress made in achieving the ultimate objective of the UNFCCC. The COP meets every year, unless the parties decide otherwise.

1.2.3.3 TCA key themes and provisions

Key themes and provisions relevant to UK construction from the TCA, including those from a House of Lords article (Coleman and Newson, 2021), are explained.

The TCA provisions include tariff-free and quota-free trade in goods, where goods meet the relevant rules of origin, although the UK has left the EU customs union and single market. A particular challenge is that different products have diverse rules of origin.

Also, companies will incur additional compliance costs to complete extra paperwork and declarations when moving goods across the EU–UK border.

1.2.3.4 TCA level playing field for open and fair competition and sustainable development

The TCA includes level-playing-field provisions to prevent distortions to trade and investment, for example, through state subsidies or reductions in environmental regulations standards (Charlson, 2021b).

Article 1.1 of Title XI: Level Playing Field for Open and Fair Competition and Sustainable Development of the TCA (HMG, 2020a, p. 179) details the principles and objectives:

> 1. The Parties recognise that trade and investment between the Union and the United Kingdom … require conditions that ensure a level playing field for open and fair competition between the Parties and that ensure that trade and investment take place in a manner conducive to sustainable development.
> 2. The Parties recognise that sustainable development encompasses economic development, social development and environmental protection, all three being interdependent and mutually reinforcing, and affirm their commitment to promote the development of international trade and investment in a way that contributes to the objective of sustainable development.
> 3. Each Party reaffirms its ambition of achieving economy-wide climate neutrality by 2050.
> 4. The Parties affirm their common understanding that their economic relationship can only deliver benefits in a mutually satisfactory way if the commitments relating to a level playing field for open and fair competition stand the test of time, by preventing distortions of trade or investment, and by contributing to sustainable development. However the Parties recognise that the purpose of this Title is not to harmonise the standards of the Parties. The Parties are determined to maintain and improve their respective high standards in the areas covered by this Title.

Where a party believes that the other has taken action that could distort trade and investment, they may be able to take 'rebalancing measures' subject to certain consultation and arbitration procedures. These measures could include the imposition of tariffs (Coleman and Newson, 2021).

1.2.4 Future EU–UK relationship

The European Union (Future Relationship) Act 2020 provides that UK law is amended as necessary to comply with the UK's obligations under the TCA (except to the extent that specific UK implementing laws are put in place). Otherwise, the TCA recognises that the parties are free to regulate as they see fit, and systems of mutual recognition were not agreed. It is expected that UK law will gradually diverge from EU law.

1.3. Legislation

Acts of Parliament are made by the UK Parliament (comprising the House of Commons and House of Lords) acting in its legislative role. Before an act, also known as a 'statute', has passed through all its stages in both houses and received royal assent it is referred to as a 'bill'. Bills may be public, private or private-member's bills, with the majority being government-sponsored public bills.

Statutes may apply to all four countries of the UK. The Scottish Parliament, the Northern Ireland Assembly and the National Assembly for Wales can only pass legislation on devolved matters that apply only to their countries.

The inspiration for new legislation may come from a number of sources, including responding to emergencies, incidents and environmental obligations.

1.3.1 Emergencies and incidents

A government may be required to respond to emergencies and incidents. Occasionally, new legislation may be required to address emergency unforeseen circumstances. For example, the Coronavirus Act 2020, having been fast-tracked through Parliament, received royal assent on 25 March 2020.

A fire which destroyed Grenfell Tower in June 2017 resulted in the death of 72 people. *Building a Safer Future – Independent Review of Building Regulations and Fire Safety* (the Hackitt Review) (Hackitt, 2018), published in May 2018, set out principles for a new regulatory framework. The Building Safety Act 2022 received royal assent on 28 April 2022. Although construction law, rather than environmental law, this is an example of a government legislative response to an incident.

1.3.2 Climate change

The Climate Change Act 2008 contains a legally binding target committing the UK to net-zero greenhouse gas emissions by 2050. In 2008, the UK was committed to an 80% reduction in greenhouse gas emissions by 2050, compared to 1990 levels. In 2019, the Climate Change Act 2008 (2050 Target Amendment) Order 2019 was passed, and this increased the UK's commitment to a 100% reduction in emissions by 2050.

Under the Climate Change Act 2008, the UK has set a series of five-year legally binding carbon budgets aimed at meeting its longer-term climate targets. The UK government has set the sixth carbon budget (2033–2037) in line with the level advised by the Climate Change Committee (DESNZ and BEIS, 2021a).

The Welsh government has separately set a legally binding target of net zero by 2050, and sub-targets for greenhouse gas emission reductions under the Environment (Wales) Act 2016.

1.3.2.1 Net Zero Strategy

The UK government published its *Net Zero Strategy* for the UK in October 2021 (DESNZ and BEIS, 2021b), setting out potential pathways to a net-zero economy in 2050. Each of these pathways involves major increases in electricity generation and hydrogen production, decarbonisation of transport, buildings, and industry, and much greater levels of energy efficiency.

At the same time, it published its *Heat and Buildings Strategy* (DESNZ and BEIS, 2021c), setting out plans for long-term building decarbonisation.

1.3.3 Environmental legislation

A government will be keen to implement its election manifesto commitments. A large proportion of environmental law and policy in the UK derives from the EU, with its implementation formerly largely monitored and enforced by EU institutions such as the European Commission.

1.3.3.1 Environment Bill 2020

The Environment Bill 2020 was introduced to implement the Conservative Party's manifesto pledge to 'protect and restore our natural environment after leaving the EU' (Smith and Priestley, 2020). In addition, the basis for many of the proposals was a series of consultations, stakeholder engagement and UK government strategies, for example, *A Green Future: Our 25 Year Plan to Improve the Environment* (HMG, 2018).

1.3.3.2 Environment Act 2021

The Environment Act 2021, which comprises 8 parts and 21 schedules, received royal assent on 9 November 2021

- Part 1: Environmental governance
- Part 2: Environmental governance: Northern Ireland
- Part 3: Waste and resource efficiency
- Part 4: Air quality and environmental recall
- Part 5: Water
- Part 6: Nature and biodiversity
- Part 7: Conservation covenants
- Part 8: Miscellaneous and general provisions.

The Act contains reforms aimed at protecting the environment now the UK has left the EU, including

- a requirement for a long-term environmental improvement plan
- the setting of various environmental targets

- incorporation of environmental principles into UK law
- a new Office for Environmental Protection to replace EU oversight functions
- extending producer responsibility schemes for waste and granting powers to tackle the inefficient use of resources in products
- introducing a 'biodiversity net gain' condition for grant of planning permission, with associated powers for new conservation covenants
- strengthening air quality legislation and increasing powers in relation to enforcement of air quality standards.

Legally binding targets are to be set by the Secretary of State on air quality, water, resource efficiency and waste reduction.

1.3.3.2.1 Policy statement on environmental principles. In addition, section 17.5 of the Environment Act 2021 provides that the Secretary of State must prepare a policy statement in accordance with the following environmental principles

(*a*) the principle that environmental protection should be integrated into the making of policies
(*b*) the principle of preventative action to avert environmental damage
(*c*) the precautionary principle, so far as relating to the environment
(*d*) the principle that environmental damage should as a priority be rectified at source, and
(*e*) the polluter pays principle.

The Secretary of State must be satisfied that the policy statement will, when it comes into effect, contribute to the improvement of environmental protection and sustainable development.

Other provisions of the Environment Act 2021 will be detailed in subsequent chapters.

1.3.4 Delegated legislation

Due to pressure on parliamentary time, Parliament may lay down a framework in an act (an 'enabling act') and then grant power to some other person or organisation to make the detailed regulations. For example, under section 1 of the Building Act 1984, the Secretary of State has the power to make regulations with respect to the design and construction of buildings. The purpose of much of the Environment Act 2021 is to serve as enabling legislation for future regulations and policy-making. The advantages of delegated legislation include speed, flexibility and expertise.

Delegated legislation can be criticised in that it vests too much power in the executive. For example, the Health Protection (Coronavirus, Restrictions) (England) Regulations 2020 were made on 26 March 2020 and came into force immediately. The Secretary of State considered that 'the restrictions and requirements imposed by these Regulations are proportionate to what they seek to achieve', and was of the opinion that, 'by reason of urgency, it is necessary to make this instrument without a draft having been laid before, and approved by a resolution of, each House of Parliament' (Charlson and Dickson, 2021).

The doctrine of sovereignty of Parliament means that legislation takes precedence over case law made by judges.

1.4. Case law

When a judge makes a decision, principles from previous cases will normally be applied. Under the English legal system, a judge is bound by earlier decisions (i.e. judicial binding precedent). In the English legal system, there is a hierarchy of courts.

- The Supreme Court of the United Kingdom, formerly the House of Lords, binds all courts beneath it by its decisions but is not bound by its own previous decisions.
- The Court of Appeal is bound by its own previous decisions and the Supreme Court, and binds all the courts below.
- The High Court is bound by the Supreme Court and the Court of Appeal, and binds all the courts below.
- The Crown Court, magistrates' courts and county courts are bound by all the courts previously mentioned but do not bind any courts.

Box 1.2 *Donoghue v Stevenson* [1932] A.C. 562

This House of Lords case was one of the most important cases of the 20th century, as it laid the foundation of the modern law of tort and the doctrine of negligence in particular.

Mrs Donoghue's friend bought her a ginger beer, which was in a bottle made of dark opaque glass. Mrs Donoghue had consumed some of the contents. When the remainder of the contents were poured, the decomposed remains of a snail floated out, causing her alleged shock and severe gastroenteritis.

Mrs Donoghue was not able to claim through breach of contract as she was not party to any contract. Therefore, she issued proceedings against Stevenson, the manufacturer. The House of Lords decided that the manufacturer owed Mrs Donoghue a duty of care.

The Technology and Construction Court, part of the Business and Property Court of the High Court of Justice, handles disputes about buildings, engineering and surveying, including environmental claims. Judges can also be responsible for interpreting legislation.

1.5. Regulators and advisers

The sanction for breach of most environmental law is prosecution in the criminal courts of an individual or company by the relevant regulator. Penalties include fines and imprisonment. Company directors and officers can be prosecuted if the criminal offence was committed with their consent or connivance or was attributable to their neglect.

Regulators can serve enforcement notices on operators requiring them to rectify breaches of environmental law. Failure to comply with enforcement notices can constitute a criminal offence. In some cases, regulators can shut down an operator's activities until the breach has been rectified.

The Regulatory Enforcement and Sanctions Act 2008 provides for civil sanctions to be used by certain regulators, instead of immediate reliance on criminal prosecution. Sanctions include

- fixed or variable monetary penalties
- compliance or restoration or stop notices
- enforcement undertakings.

The responsibilities of environmental regulators (local authorities, Defra, the Environment Agency, Natural Resources Wales and the Office for Environmental Protection) are outlined. The role of the environmental advisory bodies (Natural England and the Climate Change Committee) is explained.

1.5.1 Local authorities

On a day-to-day level, practical decision-making and enforcement in relation to environmental legislation often falls to local government officers. In particular, the following issues are dealt with by local authorities

- control of noise from premises, including noise from construction sites pursuant to sections 60 and 61 of the Control of Pollution Act 1974
- town and country planning
- public health matters
- statutory nuisance
- waste collection and disposal
- responsibility for inspecting for the existence of and identifying contaminated land within their area, and then for taking action against the person responsible for the contamination or the owner/occupier of the land.

1.5.2 Department for Environment, Food and Rural Affairs (Defra)

Defra is a ministerial department, supported by 32 agencies and public bodies, which only works directly in England. Defra is responsible for improving and protecting the environment. One of Defra's priorities is to improve the environment through cleaner air and water, minimised waste and thriving plant and terrestrial and marine wildlife (HMG, 2022a).

In January 2023, Defra (2023) published The Environmental Improvement Plan 2023 for England which is the first revision of the 25 Year Action Plan (25YEP). It builds on the 25YEP vision with a new plan setting out how Defra will work with landowners, communities and businesses to deliver each of their goals for improving the environment, matched with interim targets to measure progress.

1.5.3 Environment Agency

The Environment Agency (EA) is an executive non-departmental public body, sponsored by Defra. It was established in 1996 to protect and improve the environment. Within England, the EA is responsible for

- regulating major industry and waste
- treatment of contaminated land
- water quality and resources

- fisheries
- inland river, estuary and harbour navigations
- conservation and ecology.

The EA is also responsible for managing the risk of flooding from main rivers, reservoirs, estuaries and the sea (HMG, 2021).

1.5.3.1 EA investigations

The EA has wide powers to carry out investigations, including the power to

- require information to be provided
- gain access to premises
- obtain samples
- interview site employees
- carry out works in an emergency.

1.5.3.2 EA enforcement

The EA is responsible for enforcing the law that protects the environment, and it aims to achieve these four outcomes (EA, 2019)

- stop illegal activity from occurring or continuing
- put right environmental harm or damage, also known as 'restoration' or 'remediation'
- bring illegal activity under regulatory control, and so in compliance with the law
- punish an offender and deter future offending by the offender and others.

The EA implements its range of enforcement and sanctioning options in accordance with its policy (EA, 2022) and the Regulators' Code (BIS, 2014). Legislation specifies whether or not a breach is an offence. The EA may therefore impose civil or criminal sanctions. Before the EA decides to start a prosecution case it must

- be satisfied there is a realistic prospect of securing a conviction
- be sure it is the most appropriate enforcement action to take based on the evidence in the case, and that it is in the public interest
- consider the resulting implications and consequences.

The EA's first response is usually to give advice and guidance or to issue a warning to bring an offender into compliance, when possible. Nevertheless, the EA can and will prosecute (EA, 2019).

1.5.3.3 EA prosecutions

Examples of EA prosecutions include water and construction companies.

Southern Water was fined £90 million after pleading guilty to thousands of illegal discharges of sewage. The 2021 case, which was the largest criminal investigation in the EA's 25-year history, saw pollution offences from 16 waste water treatment works and one storm overflow brought together in one prosecution (EA, 2021).

Furthermore, Thames Water, following prosecution by the EA in 2021, was fined £4 million after a 30-hour waterfall of sewage discharge. The court heard how Thames Water had failed to carry out essential maintenance to prevent blockages in a sewer that it already knew was vulnerable to blockages. Thames Water has accrued £32.4 million in fines since 2017 for 11 cases of water pollution (Defra and EA, 2021).

The construction industry has also been prosecuted and fined. Miller Homes was fined £100 000 in 2016 for its involvement in the pollution of a watercourse (Burges Salmon, 2016). The following year, Harron Homes was fined £120 000 for failing to control silt run-off from its construction site, which illegally polluted a watercourse (EA, 2017).

1.5.4 Natural England

Natural England, an executive non-departmental public body sponsored by Defra, is the UK government's adviser for the natural environment in England. Natural England's purpose is to help conserve, enhance and manage the natural environment for the benefit of present and future generations, thereby contributing to sustainable development (HMG, 2022b). Although Natural England is not a regulator, it is an influential adviser.

1.5.5 Natural Resources Wales

Natural Resources Wales (NRW), a body sponsored by the Welsh government, was formed in April 2013 and largely took over the functions of the Countryside Council for Wales, Forestry Commission Wales and the Environment Agency in Wales, as well as certain Welsh government functions. Its general purpose is set out in section 5 of the Environment (Wales) Act 2016 (an Act of the National Assembly for Wales). The roles and responsibilities of NRW include (NRW, 2020)

- advising the Welsh government, industry and the wider public about issues relating to the environment and its natural resources
- as a regulator that prosecutes those who breach marine, forest and waste industry regulations
- designating Sites of Special Scientific Interest, Areas of Outstanding Natural Beauty and National Parks, and declaring of National Nature Reserves
- responding to some 9000 reported environmental incidents a year as a Category 1 emergency responder
- being a statutory consultee to some 9000 planning applications a year
- managing 7% of the land area of Wales, including woodlands, National Nature Reserves, water and flood defences
- being a key collaborator with the public, private and voluntary sectors providing grant aid
- monitoring the environment, commissioning and undertaking research, and being a public records body.

1.5.5.1 NRW enforcement

NRW is responsible for more than 40 different types of regulatory regime across a wide range of activities, including water discharges (surface and groundwater). Its regulatory principles (NRW, 2022) include

- act proportionately
- be consistent
- be transparent
- target enforcement action
- be accountable
- have regard to wider responsibilities.

NRW's enforcement options include stop, suspension and prohibition notices. It can serve restoration and enforcement notices. It also has the power to impose criminal sanctions and monetary and civil penalties.

1.5.6 Office for Environmental Protection

The Office for Environmental Protection (OEP), formed on 17 November 2021, has primarily replaced the European Commission's oversight function following the UK's departure from the EU. Its role is to protect and improve the environment by holding governmental and public bodies to account in England. The OEP's responsibilities are detailed in part 1, chapter 2 of the Environment Act 2021.

The principal objective of the OEP is to contribute to environmental protection and the improvement of the natural environment. The OEP is mandated to act objectively and impartially, and have regard to the need to act proportionally and transparently (section 23).

1.5.6.1 OEP strategy

The OEP has prepared a strategy (OEP, 2022) on how it intends to avoid any overlap with the Climate Change Committee and cooperate with devolved environmental governance bodies. The strategy contains an enforcement policy that places particular importance on prioritising cases that have national implications, relate to ongoing or recurrent conduct, or conduct that the OEP considers may have or has caused serious damage to the natural environment or to human health, or conduct that the OEP considers may raise a point of environmental law of general public importance (section 23).

The OEP arranged for the strategy to be published and laid before Parliament. Before preparing, revising or reviewing the strategy, the OEP must consult with such persons as it considers appropriate (section 24).

1.5.6.2 OEP guidance

The Secretary of State may issue guidance on the OEP's enforcement and policy and functions. Before issuing the guidance, the Secretary of State must prepare and lay a draft before Parliament. If either House of Parliament passes a resolution, or a committee of either House of Parliament, or a joint committee of both Houses, makes recommendations in respect of the draft guidance, the Secretary of State must produce a response and lay it before Parliament. The final guidance must be published and laid before Parliament, which it is when it comes into effect (section 25).

1.5.6.3 OEP cooperation

The OEP and the Climate Change Committee must prepare a memorandum of understanding. The memorandum must set out how they intend to cooperate with one another and avoid overlap between the exercise of their functions (section 26).

Public authorities have a duty to cooperate with the OEP and give it such reasonable assistance as it requests. If the OEP considers that a particular exercise of its functions may be relevant to the exercise of a function by a devolved environmental governance body, the OEP must consult that body (section 27).

1.5.6.4 OEP governance

Schedule 1 of the Environment Act 2021 details provisions concerning the OEP's governance, including

- appointment of executive and non-executive members
- powers, committees and delegation
- procedure, funding, and annual report and accounts
- status and independence.

1.5.7 Climate Change Committee

The Climate Change Committee (CCC) is an independent, statutory body established under the Climate Change Act 2008. The CCC's purpose is to advise the UK and devolved governments on emissions targets and to report to Parliament on progress made in reducing greenhouse gas emissions and preparing for and adapting to the impacts of climate change. The CCC's focus is on

- providing independent advice on setting and meeting carbon budgets and preparing for climate change
- monitoring progress in reducing emissions and achieving carbon budgets and targets
- conducting independent analysis into climate change science, economics and policy
- engaging with a wide range of organisations and individuals to share evidence and analysis.

The UK government and national authorities may also request specific advice from the CCC on an ad hoc basis (CCC, 2021).

References

Statutes

Building Act 1984
Building Safety Act 2022

Climate Change Act 2008
Control of Pollution Act 1974
Coronavirus Act 2020
Environment Act 2021
Environment (Wales) Act 2016
European Communities Act 1972 (repealed)
European Union (Future Relationship) Act 2020
European Union (Withdrawal) Act 2018
European Union (Withdrawal Agreement) Act 2020
Regulatory Enforcement and Sanctions Act 2008

Regulations and orders

Climate Change Act 2008 (2050 Target Amendment) Order 2019
The Health Protection (Coronavirus, Restrictions) (England) Regulations 2020

Case law

Donoghue v Stevenson [1932] A.C. 562
R v Secretary of State for Transport, ex parte Factortame Ltd (No. 2) [1991] 1 AC 603

Journals

Charlson J (2021a) Briefing: Brexit and UK construction law: past, present and future. *Proceedings of the Institution of Civil Engineers – Management, Procurement and Law* **174(1)**: 3–6.

Charlson J (2021b) Briefing: Beyond Brexit: trade and procurement implications for the UK construction industry. *Proceedings of the Institution of Civil Engineers – Management, Procurement and Law* **174(3)**: 95–98.

Charlson J and Dickson R (2021) Covid-19 and construction law: comparing the UK and Australian response. *The International Construction Law Review* **38(1)**: 5–38.

Websites

BIS (Department for Business, Innovation and Skills) (2014) *Regulators' Code*. The Stationery Office, London, UK. https://www.gov.uk/government/publications/regulators-code (accessed 09/03/2023).

Burges Salmon (2016) Developer fined £100k for building contractor's watercourse pollution. https://www.burges-salmon.com/news-and-insight/legal-updates/developer-fined-100k-for-building-contractors-watercourse-pollution (accessed 09/03/2023).

CCC (Climate Change Committee) (2021) The UK's independent adviser on tackling climate change. https://www.theccc.org.uk (accessed 09/03/2023).

Coleman C and Newson N (2021) UK–EU Trade and Cooperation Agreement. House of Lords Library, UK Parliament, London, UK. https://lordslibrary.parliament.uk/uk-eu-trade-and-cooperation-agreement (accessed 09/03/2023).

Defra (Department for Environment, Food and Rural Affairs) (2023) Environmental Improvement Plan 2023. https://www.gov.uk/government/publications/environmental-improvement-plan (accessed 07/02/2023).

Defra and EA (Department for Environment, Food and Rural Affairs and Environment Agency) (2021) Thames Water fined £4 million after 30 hour waterfall of sewage discharge. Press release. https://www.gov.uk/government/news/thames-water-fined-4-million-after-30-hour-waterfall-of-sewage-discharge (accessed 09/03/2023).

DESNZ and BEIS (Department for Energy Security and Net Zero and Department for Business, Energy and Industrial Strategy) (2021a) Guidance. Carbon budgets. https://www.gov.uk/guidance/carbon-budgets (accessed 09/03/2023).

DESNZ and BEIS (2021b) *Net Zero Strategy: Build Back Greener.* Policy paper. The Stationery Office, London, UK. https://www.gov.uk/government/publications/net-zero-strategy (accessed 09/03/2023).

DESNZ and BEIS (2021c) *Heat and Buildings Strategy.* Policy paper. The Stationery Office, London, UK. https://www.gov.uk/government/publications/heat-and-buildings-strategy (accessed 09/03/2023).

EA (Environment Agency) (2017) Harron Homes fined £120,000 over construction pollution. Press release. https://www.gov.uk/government/news/harron-homes-fined-120000-over-construction-pollution (accessed 09/03/2023).

EA (2019) *Environment Agency's Enforcement and Sanctions Policy.* https://www.gov.uk/government/publications/environment-agency-enforcement-and-sanctions-policy/environment-agency-enforcement-and-sanctions-policy (accessed 09/03/2023).

EA (2021) Record £90m fine for Southern Water following EA prosecution. Press release. https://www.gov.uk/government/news/record-90m-fine-for-southern-water-following-ea-prosecution (accessed 09/03/2023).

EA (2022) Environment Agency enforcement and sanctions policy. Policy paper. https://www.gov.uk/government/publications/environment-agency-enforcement-and-sanctions-policy/environment-agency-enforcement-and-sanctions-policy (accessed 09/03/2023).

Hackitt J Dame (2018) *Building a Safer Future. Independent Review of Building Regulations and Fire Safety: Final Report.* The Stationery Office, London, UK. https://assets.publishing.service.gov.uk/government/uploads/system/uploads/attachment_data/file/707785/Building_a_Safer_Future_-_web.pdf (accessed 09/03/2023).

HMG (His Majesty's Government) (2018) *A Green Future: Our 25 Year Plan to Improve the Environment.* The Stationery Office, London, UK. https://www.gov.uk/government/publications/25-year-environment-plan (accessed 09/03/2023).

HMG (2020a) *Trade and Cooperation Agreement between the European Union and the European Atomic Energy Community, of the One Part, and the United Kingdom of Great Britain and Northern Ireland, of the Other Part.* The Stationery Office, London, UK. https://assets.publishing.service.gov.uk/government/uploads/system/uploads/attachment_data/file/948119/EU-UK_Trade_and_Cooperation_Agreement_24.12.2020.pdf (accessed 09/03/2023).

HMG (2020b) *EU–UK Trade and Cooperation Agreement: Summary.* The Stationery Office, London, UK. https://assets.publishing.service.gov.uk/government/uploads/system/uploads/attachment_data/file/962125/TCA_SUMMARY_PDF_V1-.pdf (accessed 09/03/2023).

HMG (2021) Environment Agency. https://www.gov.uk/government/organisations/environment-agency/about (accessed 09/03/2023).

HMG (2022a) Department for Environment, Food and Rural Affairs. https://www.gov.uk/government/organisations/department-for-environment-food-rural-affairs/about (accessed 09/03/2023).

HMG (2022b) Natural England. https://www.gov.uk/government/organisations/natural-england/about (accessed 09/03/2023).

NRW (Natural Resources Wales) (2020) Our roles and responsibilities. https://naturalresources.wales/about-us/what-we-do/our-roles-and-responsibilities/?lang=en (accessed 09/03/2023).

NRW (2022) Enforcement sanctions and policy. https://naturalresources.wales/about-us/what-we-do/how-we-regulate-you/our-regulatory-responsibilities/enforcement-and-sanctions-policy/?lang=en (accessed 09/03/2023).

OEP (Office for Environmental Protection) (2022) *Office for Environmental Protection: Our Strategy.* The Stationery Office, London, UK. https://www.gov.uk/government/publications/office-for-environmental-protection-our-strategy (accessed 09/03/2023).

Smith L and Priestley S (2020) *Commons Library Analysis of the Environment Bill 2019–20.* Briefing Paper CBP 8824. House of Commons Library, UK Parliament, London, UK. https://researchbriefings.files.parliament.uk/documents/CBP-8824/CBP-8824.pdf (accessed 09/03/2023).

United Nations (1992) UN Framework Convention on Climate Change. https://treaties.un.org/pages/ViewDetailsIII.aspx?src=TREATY&mtdsg_no=XXVII-7&chapter=27&Temp=mtdsg3&clang=_en (accessed 09/03/2023).

United Nations (2015) *What is the Paris Agreement.* https://unfccc.int/process-and-meetings/the-paris-agreement/the-paris-agreement (accessed 09/03/2023).

Baker F and Charlson J
ISBN 978-0-7277-6645-8
https://doi.org/10.1680/elsc.66458.029

Chapter 2
Planning and environmental permits

Francine Baker

2.1. Introduction

This chapter focuses on the planning permission process and how planning policy and laws attempt to address climate change in conjunction with the environmental permit regime in England.

There are different environment-related assessments that may need to be carried out before planning authorities and various government bodies may approve work to start or to continue on a construction project. These include an environmental impact assessment, a habitats regulation assessment and a flood risk assessment. The first two assessments are dealt with separately in Chapter 3. However, ecological assessments may also be required by the client. These can be used in conjunction with the Future Homes Standard (HCLG, 2019), which will require all new-build domestic properties to meet higher standards of insulation and prohibit fossil fuel heating systems so that CO_2 emissions produced by new homes are 75–80% lower than those built to current standards. The standard will be implemented via changes to building regulations. Various government documents explaining the updates are available online (DLUHC, 2021a).

There are many restrictions on building and construction work in addition to planning regulations to be aware of (e.g. the Building Regulations 2010), which must also be followed. There are also rules governing the standard for the design and construction of buildings. In addition, there are various permissions that may be needed concerning covenants and 'private' rights, listed buildings, conservation areas, ancient monuments and party wall restrictions. However, as they concern construction law, property law and health and safety law, they may be referred to but they will not be dealt with in depth in this publication.

Obtaining planning permission does not concern private rights such as the right to light or a neighbour's right of way. The relevant local authority is not concerned with checking private rights such as covenants and easements, which concern property law, and these are not dealt with in this chapter. The relevant law will need to be checked separately.

2.1.1 Chapter contents – planning process

This chapter will briefly refer to the main national planning framework documents for England and Wales, noting that they are distinct policies. It will refer to the key legislation for England.

Those working on construction or engineering projects need to be aware whether an appropriate planning application has been made for planning permission to carry out the work, whether it has been accepted, when work can legally start, and when new applications need to be made. Therefore, it will be explained in this chapter why and when planning permission is required.

The key legal term in this context is 'development'. What is and is not a development needs to be explained. Some developments are allowed and are therefore called 'permitted developments', and some are not, unless the regulator (either the local authority or the Planning Inspectorate) gives permission, which is called 'planning permission'. The law regarding what is permitted and what is not permitted development has changed a number of times in recent years, and so legal advice should always be obtained to check the current position.

Sometimes, the owner of land that is used for one purpose wants to change it to another type of use (e.g. from a shop to a private residence, or from a factory to a shop). The industry needs to be aware that constructing a change of use is usually subject to planning requirements. Therefore, this chapter will refer to the laws concerning 'a change of use'. This can be a complex area of law as there are various 'use classes' and, as the content of the classes can change, legal advice is recommended.

This chapter will also discuss who may apply for planning permission, certificates of lawful use, the different types of planning application that may need to be completed depending on the type of development, and how to make the applications and where to find templates. It will discuss what are reserved matters and explain retrospective planning applications. It will consider how the planning decision is made, the grounds for appeal against refusal or a failure to consider, when there is a planning breach and what enforcement of planning breaches involves, and when the enforcement may be challenged, including judicial reviews in the High Court.

The Environment Act 2021 has brought new prerequisite requirements for developers before planning permission can be granted, and these may need to be complied with during and after the construction phase. The requirements showing a biodiversity net gain which may be accompanied by covenants to maintain biodiversity for many years will be discussed further in the context of Chapter 3, Sections 3.6.1 and 3.6.3.

2.1.2 Chapter contents – environmental permits for pollution control

Construction work may involve discharges of water or chemicals or the carrying out of flood risk activities near main rivers, the sea or other waters. Where there is the potential for pollution, an environmental permit is usually required by law. Environmental permits also promote best practice and consistency within the industry, and assist the government to achieve its environmental targets.

This chapter explains the activities that are regulated, who applies to undertake activities, where to obtain free advice about the application itself, and how to find out if an activity is exempt, as well as time limits and transfers of permits.

2.2. Planning frameworks

The National Planning Policy Framework (NPPF) was last revised July 2021 (HCLG, 2021). It is the key government document concerning the planning process in England. It provides the framework for producing local plans for housing and other development, which in turn provide the background against which applications for planning permission are decided. It states that the purpose of the planning process is to contribute to sustainable development. This concerns three overarching objectives – economic, social and environmental. The key to this framework is the 'presumption in favour of sustainable development'. Local planning authorities (LPAs) are required to comply with this framework, and hence with this presumption, when devising their local development plans and when considering planning applications. This requires, among other matters, that when an LPA is making its local plan, it should seek to improve the environment and to mitigate and adapt to the effects of climate change (HCLG, 2021, pp. 5–6).

When local authorities are devising their development plans, they must refer to the NPPF, as it is a material consideration: see section 38(6) of the Planning and Compulsory Purchase Act 2004 and section 70(2) of the Town and Country Planning Act 1990 (TCPA). The NPPF was last revised July 2021.

Although the NPPF is relevant to national infrastructure projects, it does not contain specific policies for them. These are determined in accordance with the framework in the Planning Act 2008 (as amended) and other relevant policies. Planning permission decisions are made by the Planning Inspectorate and not by the LPA.

2.2.1 National development framework for Wales

There is much environmental law that concerns both England and Wales, but, when it comes to planning law, Wales has its own planning legislative framework and its own policies. The Planning (Wales) Act 2015 introduced a new legal framework for the Welsh ministers to prepare a national land use plan, to be known as the National Development Framework for Wales (Welsh Government, 2023).

Similar to the framework set up for England, the Welsh framework sets out national land use priorities and infrastructure requirements for Wales, provides for the production of strategic development plans, covers housing supply and areas for economic growth and regeneration, provides for pre-application consultation, requires local planning authorities to provide pre-application services, and requires planning applications for nationally significant projects to be made to the Welsh ministers (Welsh Government, 2022).

2.3. Why and when is planning permission required

If you build without obtaining planning permission, you may be breaking the law. For example, if development is proposed which changes the use of a building (e.g. from residential to a commercial use), or the proposal is for constructing a new structure, demolishing a building(s), making major renovations to a property, developing an estate or any substantial construction project, you may need to obtain planning permission from the LPA.

Note that obtaining planning permission from the LPA is a separate matter from obtaining Building Regulations approval. The latter is not dealt with in this book. Both approvals may be required for building works. In addition, the applicant needs to check that the proposed work does not infringe any legal rights such as easements or covenants, a right of way or party wall restrictions.

There is a different planning permission process for nationally significant infrastructure projects (NSIPs) than there is for other building projects. Decisions are made by the Planning Inspectorate. NSIPs are dealt with in Section 2.11.

The following sections in this chapter concern planning permissions required by the LPA for the area where the proposed work is to be done.

2.3.1 Developments and planning permission

Section 57, subsection 1 of the TCPA requires that, subject to a few conditions, any person or body proposing to carry out any 'development' of land should obtain planning permission from the relevant LPA before starting work. Note that, under section 2A of the TCPA, if the proposal concerns work of potential strategic significance in the Greater London area, the Mayor of London may direct that he is the LPA for determining an application that has been sent to the LPA to obtain planning permission. Under section 74(1B) of the TCPA, the Mayor may direct an LPA to consult with him before deciding an application and before refusing an application.

Therefore, planning permission is not needed if the work is not classed as 'development'. Later, it will be discussed that planning permission is also not needed if the work is classed as 'permitted development'. (However, it is always advisable to consult with the LPA before commencing any building work.)

It is therefore necessary to know what constitutes a 'development'.

2.3.1.1 What is development?

Section 55(1) of the TCPA defines 'development' as

> the carrying out of building, engineering, mining or other operations in, on, over or under land

or

> the making of any material change in the use of any buildings or other land.

Section 55(1A) states that

> For the purposes of this act 'building operations' include
>
> (*a*) demolition of buildings
> (*b*) rebuilding
> (*c*) structural alterations of or additions to buildings, and
> (*d*) other operations normally undertaken by a person carrying on business as a builder.

However, under section 55(2) TCPA the following operations or uses of land shall not be taken to involve development of the land and therefore do not require planning permission.

(*a*) the carrying out for the maintenance, improvement or other alteration of any building of works which

(i) affect only the interior of the building, or
(ii) do not materially affect the external appearance of the building, and are not works for making good war damage or works begun after 5 December 1968 for the alteration of a building by providing additional space in it underground.

(*b*) the carrying out on land within the boundaries of a road by highway authority of any works required for the maintenance or improvement of the road [but, in the case of any such works which are not exclusively for the maintenance of the road, not including any works which may have significant adverse effects on the environment]
(*c*) the carrying out by a local authority or statutory undertakers of any works for the purpose of inspecting, repairing or renewing any sewers, mains, pipes, cables or other apparatus, including the breaking open of any street or other land for that purpose
(*d*) the use of any buildings or other land within the curtilage of a dwelling-house for any purpose incidental to the enjoyment of the dwelling-house
(*e*) the use of any land for the purposes of agriculture or forestry (including afforestation) and the use for any of those purposes of any building occupied together with land so used.

Further, a building includes 'any structure or erection, and any part of a building … but does not include plant or machinery comprised in a building' (section 336(1)).

2.3.1.2 Works which are not 'development'

Building works which concern maintenance, improvements or other alterations inside the building, or those which do not 'materially affect' the way the outside of the building looks are not classed as 'development' and therefore do not require planning permission.

This includes, for example

- internal building works
- small alterations to the outside, such as installing alarm boxes
- putting up boundary walls and fences below a certain height
- changes of use if the intended use will be incidental to the existing use(s)
- certain uses for agriculture or forestry.

There are also exceptions to having to obtain planning permission for a 'development', such as permitted development, and certain changes in the use of a building. However, even then, you should check with the local council's planning department that there are not any conditions or limitations on the permission.

2.3.2 Permitted development

Planning permission is not needed if the development is classed as 'permitted development'. This concerns a general permission that is given by the government not by an LPA.

The Town and Country Planning (General Permitted Development) (England) Order 2015 (as amended) (GPDO) states which classes of development are automatically given planning permission. It provides for the granting of certain classes of development without the requirement for a full planning application to be made.

Most home-improvement projects do not usually require planning permission. They are classed as permitted developments. However, extensions to existing houses and flats usually require planning permission.

2.3.2.1 Prior approval requirement

Permitted development rights concerning certain classes of permitted development are often subject to conditions and limitations. One condition is called the 'prior approval' requirement. It requires the applicant to apply to the LPA for it to consider whether prior approval will be required.

Developments such as those in a conservation area or in some cases where an environmental impact assessment is required will not have permitted development rights, neither will developments where such rights are restricted by planning conditions or obligations.

Each category of permitted development or of a change of use, for example, concerning protected buildings or listed buildings, is subject to various restrictions and designations. There may also be limits on the size of the proposed change of use. Therefore, the latest updated GPDO (i.e. at present, as amended by the 2021 Order) should be checked for each proposal.

Where prior approval is required under the GPDO, the operator will not need to submit a full planning application to the LPA but will be required to submit an application for 'prior approval'. This allows the LPA to consider the proposals, their likely impacts and how these can be mitigated.

2.3.2.2 When to apply for a lawful development certificate (LDC)

Whether or not a proposed project qualifies as permitted development is not always clear. In such cases it is advisable to apply for a lawful development certificate (LDC) from the LPA (HMG, 2023f).

If an applicant is unsure as to whether an activity or building operation will be considered lawful for planning purposes, an application may be made to the LPA for a certificate of lawful use (CLU) under section 191 or section 192 of the TCPA. The certificate may be granted under section 191 for an existing use or development, or under section 192 for a proposed use or development. The certificate is not a planning permission. It provides evidence that the building work is lawful.

Under section 39 of the Town and Country Planning (Development Management Procedure) (England) Order 2015 the prescribed fee for the application must be paid and the application must describe precisely what is being applied for.

For example, if it is an application under section 191 the following information should be included to support the application

- clearly dated photographic evidence
- statutory declarations from applicants, former owners and neighbours
- council tax or electoral records
- relevant invoices or receipts concerning the use of the land or the operations carried out there.

If it is an application under section 192 then it only concerns the future use or future actions, and whether the intended use or actions would be lawful.

In relation to both types of application, it is an offence to knowingly or recklessly make a materially false or misleading statement.

An LPA can remove some of your permitted development rights by issuing an Article 4 direction. This is issued where the character of an area may be threatened. It usually concerns conservation areas.

An applicant should check with the LPA whether there is any such direction in place before starting work under permitted development rights.

2.3.2.3 The Community Infrastructure Levy (CIL)

Although a development may have permitted rights, it may still be subject to a levy. The Community Infrastructure Levy (CIL) is imposed by part 11 of the Planning Act 2008. Liability for the CIL may arise where there is a charging schedule in the relevant LPA area and once planning permission has been granted or deemed to have been granted.

It is good practice to check with and consult with the LPA before starting work.

2.3.2.4 Changes to permitted development rights

Many changes have been made to the GPDO. The Town and Country Planning (General Permitted Development) (England) (Amendment) Regulations 2020 were made on 9 November 2020 (the 2020 Regulations); they came completely into force on 6 April 2021.

2.3.2.4.1 Minimum size dwelling-houses. Article 3 of the 2020 Regulations now requires all new dwelling-houses constructed under permitted development to be a minimum size of

37 square metres and to satisfy the nationally described space standards set out in 'Technical housing standards – nationally described space standard' (DLUHC, 2015).

The 2020 Regulations were introduced because builders had been constructing houses that were too small. This requirement applies to applications for prior approval submitted on or after 27 March 2015.

2.3.2.4.2 Construction of new flats in airspace. The Town and Country Planning (Permitted Development and Miscellaneous Amendments) (England) (Coronavirus) Regulations 2020 provided for a new permitted development right under Class A from 1 August 2020 (MHGLC, 2020). It allows the construction of up to two extra storeys of flats on the top residential storey of an existing purpose-built, detached block of flats building which is three storeys or more from the ground. This is subject to a maximum height for the extended storeys of 30 metres. A full planning application is not required.

The above right does not apply if the building is a listed building or a scheduled monument, or to land within their curtilages. However, always check with the LPA for guidance, as it has authority to suspend certain permitted development rights under Article 4 of the GPDO.

From 31 August 2020, new planning development rules for the construction and for upward extensions of existing homes and of new homes above a building came into force under the Town and Country Planning (General Permitted Development) (England) (Amendment) (No. 2) Order 2020 (the 2020 Order No. 2). However, this Order contains conditions and limitations on how the new permitted development rights will operate. The developments will require prior permission from the relevant LPA.

2.3.2.4.3 Mobile networks. The Town and Country Planning (General Permitted Development) (England) (Amendment) Order 2022 came into force on 4 April 2022. The changes in this Order concern mobile network development. It provides 'permitted development rights' for electronic communications infrastructure.

The aim is to ensure that the planning system addresses the disparities in digital connectivity between rural and urban areas, so that, by 2030, the UK will have nationwide gigabit-capable broadband and 4G coverage, with 5G coverage for the majority of the population.

The legislation amends Class A of part 16 of schedule 2 to the GPDO. The changes came into force on 4 April 2022 and apply to England and Wales only. These new planning conditions amend part 16 of the GPDO so that the following is required

1. Code Operators are now required to minimise the visual impact of new development on the surrounding area as far as possible, particularly considering potential impacts on Article 2(3) land when installing equipment.

Article 2(3) land is stated in part 1 of the GPDO as (words in italics added by author)

PART 1

Article 2(3) land

1. Land within

   (*a*) an area designated as a conservation area under section 69 of the Planning (Listed Buildings and Conservation Areas) Act 1990 (designation of conservation areas)
   (*b*) an area of outstanding natural beauty
   (*c*) an area specified by the Secretary of State for the purposes of section 41(3) of the Wildlife and Countryside Act 1981 ([*which Secretary of State specifies will be the subject of*] enhancement and protection of the natural beauty and amenity of the countryside)
   (*d*) the Broads
   (*e*) a National Park, and
   (*f*) a World Heritage Site.

2. Operators must consider and minimise impacts on the accessibility of footways and access to properties.
3. Single developments of radio equipment housing up to 2.5 cubic metres in volume are permitted without prior approval on Article 2(3) land. (A single development means each unit of equipment housing, rather than all housing deployed at one time.) Greater volumes are subject to prior approval.
4. The volume limits on radio equipment housing where it is located in a compound are no longer applied.
5. Existing ground-based masts less than a metre wide are replaced or altered with increases in width up to two-thirds without the need for prior approval in all areas.
6. Greater increases in width are subject to prior approval. If the ground-based masts are one metre or greater in width, the alteration or replacement of the mast with increases in width up to one-half or two metres (whichever is greater) without the need for prior approval in all areas.
7. Alteration or replacement of existing ground-based masts which would increase the height up to 25 metres is permitted, subject to prior approval on Article 2(3) land or land on a highway and land on or within SSSI.
8. Greater increases in height up to 30 metres are subject to prior approval.
9. Increases in height should be calculated by comparing width at the widest parts of the existing and new masts.

10. Provision for new planning conditions to be included in part 16 of the GPDO to provide added protection for more sensitive areas where development may not require prior approval.
11. Building-based masts.
12. On buildings which are less than 15 metres in height, the installation, alteration or replacement of building-based masts up to 10 metres in height above the tallest part of the building within 20 metres of the highway is permitted subject to prior approval.
13. Prior approval is no longer needed for the installation, alteration or replacement of building-based masts up to 6 metres in height above the tallest part of the building.
14. These changes will only apply on unprotected land, that is, land that is not Article 2(3) land or SSSI (a Site of Special Scientific Interest). It describes an area that's of particular interest to science due to the rare species of fauna or flora it contains or important geological or physiological features. The existing conditions which limit the height of masts and require visual impact to be minimised on buildings will continue to apply.
15. New ground-based masts.
16. The installation of new ground-based masts up to 25 metres on Article 2(3) land or land on a highway, and up to 30 metres on all other land (except land on or within SSSI), is permitted subject to prior approval.
17. All new masts that exceed these heights will require full planning permission.

2.3.2.4.4 Code of practice for wireless network. A new Code of Practice for Wireless Network Development in England provides updated guidance for all those involved in network deployment. It focuses on the siting and design of wireless infrastructure and the process for engaging with local authorities and communities (DCMS and DSIT, 2022).

2.3.3 Development exceeds GPDO limitations

Where the principle of development is not an issue, the LPA can only consider the siting and appearance of the proposal. Where the development exceeds the limits identified in the GPDO, full planning permission will be required, and operators will need to submit a planning application to the LPA in the ordinary manner.

2.3.4 Demolition and planning applications

Any building operation consisting of the demolition of a building was permitted under Class B of part 11, schedule 2 of the 2015 GPDO. However, the demolition of a concert hall, a live music venue or a theatre must now be considered as part of a planning application (Regulation 6 of the 2020 Regulations).

The demolition of undesignated statues and monuments (i.e. those that are not already listed structures) is not allowed without the permission of the LPA.

2.3.4.1 Demolition and rebuild permissions

From 31 August 2020, under the 2020 Order No. 2, a planning application to demolish and rebuild vacant and redundant residential and commercial buildings is not required if the buildings are being rebuilt as homes. This applies to vacant and redundant free-standing buildings that fell within the following use classes on 12 March 2020.

B1(a) offices

B1(b) research and development

B1(c) industrial processes (light industrial), and free-standing purpose-built residential blocks of flats (C3).

However, the building must have been vacant for at least six full months prior to the date of the application for prior approval and built before 1 January 1990.

This new permitted development right allows for redevelopment of one new building within the footprint of buildings with a footprint of up to but no greater than 1000 square metres, and with a maximum height of 18 metres.

2.4. Change of use of the building and use classes

If building work involves changing the use of a building to a different one, the applicant should always check whether the proposed building work is compliant with the Building Regulations, and check with the LPA whether the proposed change of use requires a planning permission application. The work will usually need planning permission if the use of the structure is in one class and what it is to be changed to is in a different use class. For example, planning permission is required to be obtained from the LPA for building operations where the GPDO does not permit it (e.g. to remove a shopfront (Class E) to convert a shop to a dwelling (Class C)).

2.4.1 Governance

The Town and Country Planning (Use Classes) Order 1987 (Use Classes Order 1987) categorises the use of land or buildings into certain use classes. This Order was amended (changed) by the Town and Country Planning (Use Classes) (Amendment) (England) Regulations 2020 (the 2020 Use Regulations), which came into force on 1 September 2020.

Classes A, B and D in the Use Classes Order 1987 were revoked by the 2020 Use Regulations.

2.4.2 Recent changes

From 21 April 2021, Class B proposals to demolish commemorative unlisted statues, memorials and monuments now require full planning permission before work can legally start.

However, always check with the relevant LPA, as planning law classes may change.

The new (2021) use classes are set out below.

Class B

B2 General industrial – Use for industrial process other than one falling within Class E(g) (previously Class B1) (excluding incineration purposes, chemical treatment, landfill or hazardous waste).
B8 Storage or distribution – This class includes open air storage.

Class C

C1 *Hotels* – Hotels, boarding and guest houses where no significant element of care is provided (excludes hostels).
C2 *Residential institutions* – Residential care homes, hospitals, nursing homes, boarding schools, residential colleges and training centres.

C2A Secure residential institution – Use for a provision of secure residential accommodation, including use as a prison, young offenders' institution, detention centre, secure training centre, custody centre, short-term holding centre, secure hospital, secure local authority accommodation or use as a military barracks.

C3 *Dwelling-houses* – This class is formed of three parts.

C3(a) use by a single person or a family (a couple whether married or not, a person related to one another with members of the family of one of the couple to be treated as members of the family of the other), an employer and certain domestic employees (such as an au pair, nanny, nurse, governess, servant, chauffeur, gardener, secretary and personal assistant), a carer and the person receiving the care and a foster parent and foster child
C3(b) use by up to six people living together as a single household and receiving care (e.g. supported housing schemes such as those for people with learning disabilities or mental health problems)
C3(c) use by groups of people (up to six) living together as a single household. This allows for those groups that do not fall within the C4 definition of a house in multiple occupation, but which fell within the previous C3 use class, to be provided for (i.e. a small religious community may fall into this class, as could a homeowner who is living with a lodger).

C4 *Houses in multiple occupation* – Small shared houses occupied by between three and six unrelated individuals, as their only or main residence, who share basic amenities such as a kitchen or bathroom.

Class E – *Commercial, business and service*

In 11 parts, Class E more broadly covers uses previously defined in the revoked Classes A1/2/3, B1, D1(a–b) and 'indoor sport' previously defined in revoked Class D2(e)

E(a) Display or retail sale of goods, other than hot food.
E(b) Sale of food and drink for consumption (mostly) on the premises.
E(c) Provision of

E(c)(i) Financial services
E(c)(ii) Professional services (other than health or medical services)
E(c)(iii) Other appropriate services in a commercial, business or service locality.

E(d) Indoor sport, recreation or fitness (not involving motorised vehicles or firearms or use as a swimming pool or skating rink).
E(e) Provision of medical or health services (except the use of premises attached to the residence of the consultant or practitioner).
E(f) Creche, day nursery or day centre (not including a residential use).
E(g) Uses which can be carried out in a residential area without detriment to its amenity.

E(g)(i) Offices to carry out any operational or administrative functions.
E(g)(ii) Research and development of products or processes.
E(g)(iii) Industrial processes.

Class F – *Local community and learning*

In two main parts, Class F covers uses previously defined in the revoked Classes D1 and D2(e) ('outdoor sport', 'swimming pools' and 'skating rinks'), as well as newly defined local community uses.

F1 *Learning and non-residential institutions* – Use (not including residential use) defined in seven parts.

F1(a) Provision of education.
F1(b) Display of works of art (otherwise than for sale or hire).
F1(c) Museums.
F1(d) Public libraries or public reading rooms.
F1(e) Public halls or exhibition halls.
F1(f) Public worship or religious instruction (or in connection with such use).
F1(g) Law courts.

F2 *Local community* – Use defined in four parts.

F2(a) Shops (mostly) selling essential goods, including food, where the shop's premises do not exceed 280 m^2 and there is no other such facility within 1000 m.
F2(b) Halls or meeting places for the principal use of the local community.
F2(c) Areas or places for outdoor sport or recreation (not involving motorised vehicles or firearms).
F2(d) Indoor or outdoor swimming pools or skating rinks.

Sui generis

Sui generis is a Latin term that, in this context, means 'in a class of its own'. Certain uses are specifically defined and excluded from classification by legislation, and therefore become *sui generis*. These are

- theatres
- amusement arcades/centres or funfairs
- launderettes
- fuel stations
- hiring, selling and/or displaying motor vehicles
- taxi businesses
- scrap yards, or a yard for the storage/distribution of minerals and/or the breaking of motor vehicles
- 'alkali work' (any work registerable under the Alkali, etc. Works Regulation Act 1906 (as amended))
- hostels (providing no significant element of care)
- waste disposal installations for the incineration, chemical treatment or landfill of hazardous waste
- retail warehouse clubs
- nightclubs
- casinos
- betting offices/shops
- pay day loan shops
- public houses, wine bars or drinking establishments – from 1 September 2020, previously Class A4 drinking establishments with expanded food provision; from 1 September 2020, previously Class A4 hot food takeaways (for the sale of hot food where consumption of that food is mostly undertaken off the premises); from 1 September 2020, previously Class A5
- venues for live music performance – these are defined as *sui generis* use from 1 September 2020
- cinemas – from 1 September 2020, previously Class D2(a)
- concert halls – from 1 September 2020, previously Class D2(b)
- bingo halls – from 1 September 2020, previously Class D2(c)
- dance halls – from 1 September 2020, previously Class D2(d).

Other uses become *sui generis* where they fall outside the defined limits of any other use class. For example, C4 (houses in multiple occupation) is limited to houses with no more than six residents. Therefore, houses in multiple occupation with more than six residents become a *sui generis* use.

2.4.3 Changes of use

The government attempted to consolidate and simplify permitted development rights in England in the Town and Country Planning (General Permitted Development etc.) (England)

(Amendment) (No. 2) Order 2021 published on 9 July 2021 (the 2021 Order). This Order came into effect on 1 August 2021. However, in the few cases where proposals are no longer a permitted development under the 2021 Order, the old law applied to those proposals until 31 July 2022. Such cases have been classified as 'protected development'.

The main purpose of the 2021 Order is to amend the GPDO to reflect the changes made to the Use Classes Order 1987 by the 2020 Use Regulations. Most of the changes concern part 3 (Changes of use) and part 4 (Temporary buildings and uses). These parts grant permitted development rights for certain changes of use; for example, a change of use from a hotel (C1) to a state-funded school (F1A), or from small houses in multiple occupation (C4) to dwelling-houses (C3).

Part 3, Class R remains unchanged. It allows a change of use of an agricultural building to a 'flexible use' falling within Class E. A change of use of agricultural buildings to the former D2 uses, such as a cinema or live music venue, is no longer permitted under the GPDO due to the 2021 Order. However, a planning application may be made, as is the case for any changes of use that were not granted permitted development rights in the GPDO.

Again, this is an area of planning law that may change, so always check with the LPA.

2.4.3.1 New Class MA

Probably the most significant change to the GPDO (the 2021 Order) is the new Class MA, which allows the conversion of various commercial uses into dwellings without the need for express planning permission. From 1 August 2021, an application for prior approval needs to be made before any development is carried out. The applicant will have to show that the building

- has been vacant for three months prior to the application
- has been in commercial use for at least two years
- will not be subject to more than 1500 square metres of floorspace change.

The range of eligible uses are set out within the regulations.

Like other permitted development rights under the GPDO, this change does not apply to listed buildings or to locations within an Area of Natural Beauty or a National Park (Planning Portal, 2023a).

The Planning Portal provides information about the various permanent or temporary changes of use (Planning Portal, 2023b).

2.5. Who can apply for planning permission?

It is the property owner's responsibility to seek planning permission, although they may do this through a legally authorised agent. Always contact the LPA if you are in any doubt, and keep an accurate record of their advice, as you should do with all advice received from the LPA.

2.5.1 Recent requirements
From 6 April 2021, all new homes delivered through permitted development are also required to meet nationally described space standards. An application form asking for the LPA's prior approval of certain proposals, such as regarding the new Class G or Class MA, must also be submitted and approved before work can legally commence.

As a result of the 2021 Order, from 1 August 2021, before work can begin the LPA must assess proposals in relation to health and environmental considerations. Namely

- the risks of contamination
- flood risk
- impacts of noise from commercial premises
- natural light to all habitable rooms
- the storage of domestic waste.

The default position is that, after submission of a prior approval application, the LPA must make a decision within 8 weeks. This is subject to agreed extensions of time for approval and any limitations of the application. If the timeframe is not complied with, the applicant may appeal the decision.

2.6. Planning applications
There are three main types of planning application – outline, reserved matters and full – and these will be discussed in this section. However, there is a range of other types of application, and these are referred to in Sections 2.6.8 and 2.6.9. Template planning application forms are covered in Section 2.6.10.1.

2.6.1 Pre-application enquiries
Potential applicants can make enquiries of the LPA before submitting a formal planning application. LPAs have no statutory duty to respond to such enquiries, but they do have the power to do so. The applicant may request a meeting with the LPA's planning officer. The advice obtained from the LPA can help avoid submitting an invalid application.

The applicant should provide sufficient information to enable an informed response. They should discuss any site problems, such as roads, or sewers, power cables, the location of utilities, or other issues such as noise and traffic, and ask whether the LPA may be likely to impose conditions on the granting of planning permission rather than reject an application.

If your project is not in accordance with the development plan for the area you will need to justify to the LPA or planning officer (if at a meeting) why your project should go ahead.

2.6.1.1 Costs
Under section 93 of the Local Government Act 2003 LPAs can charge for responding to a pre-application enquiry. However, the power to charge under the Act is limited to cost recovery. It does not include a power to make a profit. How much a particular authority may charge will

vary depending on the advice sought. The UK government advises that LPAs should be transparent by providing clear online information such as

- the scale of charges for pre-application services applicable to different types of application (e.g. minor, major or other)
- the level of service that will be provided for the charge, including
  - the scope of work and what is included (e.g. duration and number of meetings or site visits)
  - the amount of officer time to be provided (recognising that some proposed development requires input from officers across the local authority, or from other statutory and non-statutory bodies)
  - the outputs that can be expected (e.g. a letter or report) and firm response times for arranging meetings and providing these outputs.

LPAs may provide links to any charges that statutory consultees such as a parish council or a Neighbourhood Forum may levy for pre-application advice, where this is known.

2.6.2 LPA responsibility

LPAs should make it clear to applicants that they do not owe a duty of care regarding the provision of their pre-action advice, and that therefore they are not bound by their pre-application advice. If the applicant claims that pre-action advice was given by the LPA negligently, the remedy is a claim to the local commissioner for administration (the local ombudsman) for maladministration.

The nature and location of particular schemes may justify other parties being involved in the response to this claim, such as elected members of the LPA (section 25 of the Localism Act 2011) and statutory consultees.

2.6.3 Planning application fees

The cost of making a planning application can be calculated using the calculator on the Planning Portal (Planning Portal, 2023c).

For example, at the time of writing, the fee for a full planning application for 50 new dwelling-houses is £23 100.00, assuming no exemptions apply, whereas the fee for an outline planning application concerns the site area – for a site of 2000 square metres the fee is only £924, assuming there are no exemptions.

2.6.4 Planning and flood risk assessments

A flood risk assessment may be needed before the LPA will grant planning permission for the development to go ahead. These assessments should usually be completed by a professional flood risk assessor at the design stage of the proposed development, so that any changes can be incorporated in any architectural and engineering designs.

If the development is within a flood zone, a flood risk assessment will be needed as part of the planning application. However, not all development proposals require a flood risk assessment for planning consent. For example, an extension to a house does not usually require a flood risk assessment, but it depends on the individual project (DLUHC, 2022a). More information related to flood risks is given at 2.12.6 and in Chapter 3 at 3.3.7.4.

Information about short and the long-term flooding in an area, including river, sea, groundwater and rainfall levels, can be obtained from the government's online service or by calling the government offices in England, Wales, Scotland or Northern Ireland (HMG, 2023a, 2023g).

The government webpages provide useful information and a service so that you can find out if a flood risk assessment is needed as part of your planning application (HMG, 2021b). You can also download a printable flood zone map.

2.6.5 Outline planning application

An outline planning application is used to find out whether the LPA is likely to give permission before a full planning application is made, and therefore before substantial full planning application fees and costs have been incurred.

An outline application is appropriate where the applicant wants to ask whether the 'principle of development' is acceptable to the LPA, without having to supply expensive detailed plans and other documents. The application needs to provide only general information about the project. Therefore, an outline application is often used by developers of large, complex or expensive schemes in order to save costs.

The completed outline application form, a plan and a fee should be presented to the relevant LPA. LPAs have the power to decline to consider an application in outline form. Outline applications for development in a conservation area may not be accepted.

2.6.5.1 When can work start?

A successful outline planning application does not provide permission to start work on site. The outline permission usually has conditions attached, such as requiring the later approval of 'reserved matters'. The latter refers to information that has not been included in the initial outline planning application.

An application concerning 'reserved matters' under section 92 of the TCPA must be made within three years of receiving approval of the outline application. The outline permission notice will state which matters have been reserved for later approval.

The development should start no later than two years from the final approval of the reserved matters (TCPA section 92). The LPA does have the power, however, to substitute other time limits if it wishes (TCPA 1990 section 92(4)).

If work has not started within the time limit allowed by the permission, the applicant will need to submit a fresh application if they wish to undertake that development.

2.6.6 What are 'reserved' matters?'

The benefits of obtaining an outline application before a making full planning application are that the developer/client can find out it if the development will be accepted in principle before any extra costs are incurred. A planning application to deal with reserved matters must be made within three years after the outline planning application is approved.

Reserved matters concern details of the proposed development that were not included in the outline application. They are defined in the Town and Country Planning (Development Management Procedure) (England) Order 2015 (as amended). They comprise detailed drawings showing

- access
- appearance
- landscaping
- layout
- scale (within upper and lower limits for the height, width and length of each building).

Landscaping is given a broader meaning than before and new definitions of 'access', 'appearance', 'layout' and 'scale' were first introduced in the Town and Country Planning (Development Management Procedure) (England) Order 2010.

2.6.6.1 Applications for approval of reserved matters

It is very important to submit the correct application when applying for approval of reserved matters. If an application form for full planning consent is submitted by mistake, the LPA has the power to consider it as if it were a completely fresh application!

The application for reserved matters should

- be in writing
- contain details of the outline planning permission
- include such plans and drawings as are necessary to deal with the reserved matters
- be accompanied by the correct fee (otherwise the application will not be registered).

Once outline permission has been granted, the LPA can only make a refusal of an application for approval of reserved matters on the basis that the details of the reserved matters are not satisfactory.

The substance of the reserved matters application should not differ substantially from the development proposed in the outline application. If it does, the LPA will be entitled to consider it as a fresh application for planning permission. This could lead to a refusal.

The time allowed for an LPA to approve a reserved matters application is usually eight weeks for a minor development and 13 weeks for a major development. However, these times may vary depending on the complexity and details of the project.

2.6.6.2 LPA's refusal of reserved matters application

In granting an outline permission, the LPA has approved the 'principle of development'. The LPA cannot subsequently use the application for reserved matters as an opportunity to refuse consent on the grounds that the principle of the development is unacceptable. A refusal on the principle of whether or not development should be allowed is not possible at this stage.

However, if a situation arises where, for example, an environmental impact assessment is required at the reserved matters stage (where none was identified as necessary at the outline stage), then the principle of whether planning permission should be granted may need to be reconsidered.

2.6.7 Full planning consent applications (section 62 of the TCPA)

Section 62 of the TCPA sets out

- the form and manner in which the full planning application must be made
- the particulars of matters that are to be included in the application
- the documents or other materials that are to accompany the application.

A full planning consent application is appropriate

- where permission is required for a material change of use
- where temporary planning permission is sought
- for detailed consideration of a new development (not just the principle of it).

The following must be submitted (usually with three copies where appropriate)

- a completed application form
- a plan describing the subject matter of the application
- the relevant fee.

A full planning consent application is also required for the subdivision of a house or for any works relating to a flat, to changes in the number of dwellings, for the conversion of flats and for changes in the use of all or part of a property to a non-residential use.

2.6.8 The Householder Planning Consent Application Form

This form is used for works to alter or enlarge a single house or garden or works within the house's curtilage (boundaries); for example, for loft extensions, conservatories, extensions to the house, garages, outbuildings, swimming pools, walls and fences, and any access for vehicles, including footway crossovers.

2.6.9 Other applications

There are various types of consent that may need to be obtained before work can legally commence. This may be in addition to a planning application or a separate planning process. For example, the following types of consent may need to be obtained before work can commence.

- **Listed building consent.** This consent cannot be applied for online. It is obtained from Historic England, the object being to protect the unique historical or architectural character of the building or structure and to keep a register of listed buildings (Historic England, 2023a). An application for listed building consent for alteration, extension or demolition of a listed building is made under the Planning (Listed Buildings and Conservation Areas) Act 1990. An application should be submitted for all planning applications requiring alteration, extension, or demolition to a listed building (Planning Portal, 2023d).
- **Scheduled monument consent.** This consent cannot be applied for online. It is obtained from Historic England, the object being to protect the nationally important historical, archaeological, architectural, artistic or traditional sites of interest (Historic England, 2023b). Consent must be applied for in advance of any work that would demolish, damage, remove, repair, alter or add to a scheduled monument, or of carrying out any flooding or tipping operations on land in, on or under a monument (HMG, 2023c; Planning Portal, 2023e). A schedule of monuments (the list of legally protected monuments) is maintained under section 1 of the Ancient Monuments and Archaeological Areas Act 1979.
- **Advertisement consent.** This concerns proposals to display an advertisement or sign (Planning Portal, 2023f). The definition of an advertisement is broadly defined under section 336(1) of the TCPA (as amended). The display of advertisements is subject to a separate consent process within the planning system, which is principally set out in the Town and Country Planning (Control of Advertisements) (England) Regulations 2007 and no separate planning permission is required in addition to advertisement consent (DLUHC, 2019).
- **Tree preservation order consent.** LPAs make these orders to protect the amenity of trees and woodlands in their area. The law on Tree Preservation Orders is contained in part VIII of the TCPA (as amended) and in the Town and Country Planning (Tree Preservation) (England) Regulations 2012. If any work on trees is being planned then the local LPA should be contacted to check if there is a preservation order or if the tree is in a conservation area, and to obtain consent to the work, then to do any work on the trees (DLUHC, 2014).
- **Hazardous substances consent.** A proposal for development which needs to use or store hazardous substances such as chlorine, hydrogen, or natural gas, at or above certain thresholds, requires a hazardous substances consent (HSC) before it can operate. The LPA is usually the Hazardous Substances Authority responsible for deciding whether to grant or revoke HSC. Further advice and resources are available on the Planning Portal (2023g).

2.6.10 How to find and make the planning application

It is advisable to contact the relevant LPA for advice about which application is suitable for your development. You can obtain most LPA forms for different types of application (outline, full, reserved and others) via the Planning Portal. Forms can be submitted online through the Portal or printed out and posted (Planning Portal, 2023h).

The online Planning Portal service asks you to state the relevant LPA. If the Portal does not have the form(s) for that LPA, you should contact the LPA directly.

The application for planning permission form should be used for making a full planning consent application for development (under section 62 of the TCPA 1990 (as amended)), but not for householder developments. The form asks for specific information to be submitted to make a valid application.

2.6.10.1 Template forms for planning applications

There are government application form templates for LPAs, and a template fire statement for use by applicants can be found on the UK government webpages (DLUHC, 2021b).

There are separate application form templates for the demolition of an unlisted building and for the demolition of a listed building, for householder extensions and demolitions, and a range of other application forms.

LPAs refer to these government templates to produce their own versions. For more information about how to submit a planning application, you should contact the LPA(s) where the development is taking place.

2.6.10.2 Local lists of information

The introduction of a national form and information requirements led some individual LPAs to be concerned about how they could lawfully seek additional information, if they considered that such information was essential before an application could be assessed. As a result, many LPAs produced their own 'local list' of additional information required to accompany an application. However, this in turn led to concerns that LPAs were demanding unnecessary information.

The government sought to resolve this matter in section 6 of the Growth and Infrastructure Act 2013, which requires LPA requests for information to be

> (*a*) reasonable having regard [...] to the nature and scale of the proposed development, and
> (*b*) [*provided that*] only if it is reasonable to think that [*it*] will be a material consideration in the determination of the application.

In the event that an application is not validated, the applicant can appeal to the Planning Inspectorate.

2.6.10.3 Validation and registration of application

Once the planning form, associated documentation (if any) and the planning fee have been received, the LPA will validate the submission and, if all is in order, register the application. Once an application has been registered, 'the clock starts ticking' in terms of the applicant's right to appeal on grounds of non-determination and the LPA's entitlement to government grants (for determining *X*% of applications within *Y* weeks).

Section 69 of the TCPA 1990 provides that every planning application must be entered on a register. These are kept at the offices of the LPA and are available for public inspection. The LPA then gives the applicant a written notification of registration.

2.7. How is the planning decision made?

When applying for permission for a development, the applicant should consider whether the proposed development is consistent with the relevant LPA's development plan. This plan is also called the 'local plan'. It comprises a group of planning documents that set out the LPA's policies and proposals for the use of the land in its area.

When an LPA is devising its development plan, it must refer to the National Planning Policy Framework, as this is a material consideration: The aim of a local plan should be to guide the future sustainable growth and development of the area (e.g. concerning housing, employment, leisure and retail). A development plan should also identify areas for environmental or heritage protection.

2.7.1 Consultation

Once an application has been validated and registered by the LPA there is a period of consultation, usually up to 21 days. The LPA will publish the application; usually it is placed on its website and advertised in the local newspaper. The LPA should notify relevant bodies, such as local communities, so that the public may express its views (Article 15 of the Town and Country Planning (Development Management Procedure) (England) Order 2015) as well as the Environment Agency. A closing date for any views to be received by the LPA will be given in the publicity documents. This date is 14 days following a newspaper advert, 18 days regarding a public service infrastructure development, and otherwise usually 21 days following the publicity (DLUHC, 2022b).

The LPA will assess the relevance of the various views received and may suggest amendments to the application as a result.

2.7.2 Report and decision

Then the LPA may give a report about the application to a planning committee or to a senior planning officer to decide the application. The applicant is entitled to have a copy of all relevant background papers and comments, as well as the report.

The applicant will receive a letter stating the LPA's decision, either granting or refusing permission, or granting permission subject to conditions.

2.8. Appeals

An appeal against the LPA's decision can be made to the Planning Inspectorate (Planning Inspectorate, 2022a). The appeal must be made by the planning application applicant. However, the applicant should first discuss the application with the LPA. It may be that an amended application can be submitted without extra cost, and that it will be approved.

2.8.1 Grounds for appeal

The person who made the application has a right of appeal to the Planning Inspectorate if the LPA's decision is not made within eight weeks of the application, or within six weeks if it is for a listed building (unless the period has been extended by agreement with the applicant). Other people may comment on the appeal if they do so within six weeks from the appeal start date.

Applications are made online to the Planning Inspectorate, and copies of all documents should be sent to the LPA.

2.8.2 Fee and deadlines

If the LPA rejects the application for a certificate of lawful use (CLU), the applicant can appeal against the decision if they disagree with it.

There is no fee for appealing or commenting. There is no deadline for the applicant to appeal the LPA's refusal to provide a certificate, unless it concerns a listed building, in which case the appeal must made be within six months.

Box 2.1 APP/B2355/X/20/3257581 – Installation of underground pipe was a 'material operation'

An appeal was made by Mr I. Hunt under section 195 of the Town and Country Planning Act 1990 (TCPA) as amended by the Planning and Compensation Act 1991 against the refusal of Rossendale Borough Council to grant a certificate of lawful use or development.

The key issue was whether the works that had been carried out constituted the commencement of development. The planning inspector did not consider that the digging of a trench (approximately 5 metres long and 1 metre deep) and the laying of an underground plastic drainage pipe could be classified as *de minimis* works. Rather, the inspector considered that the works constituted a 'material operation' under section 56(4)(c) of the TCPA 1990, and that the definition of what constitutes a start of development as set out in the Act was satisfied. Therefore, the development in question had been lawfully commenced.

It was therefore held that Rossendale Borough Council's refusal to grant a certificate of lawful use or development was not well founded and that the appeal should succeed.

If the application is refused, is subject to conditions the applicant considers unacceptable or is not decided within the statutory time limits, or an application for the approval of reserved matters is not approved, the applicant may appeal to the Secretary of State (section 78 of the TCPA 1990). However, this should be a last resort.

2.8.3 Time limits

Most appeal applications with all relevant documents must be received by the Secretary of State within six months of the date of the LPA's decision. Householder application appeals must be filed within 12 weeks.

If the authority's decision was not made within eight weeks, then the period of six months starts after the decision was due to be made.

2.8.4 How to appeal – the application

Appeal can be made online or by post, but the latter can take much longer.

The following documents should be supplied as part of your appeal

- the original planning application
- the LPA decision – if there was no decision, then the letter that acknowledged receipt of the application
- all documents that were sent to the LPA
- any other documents supporting the appeal, including a map of the surrounding area
- a full statement of the appeal case.

A copy of the appeal and all relevant documents must also be sent to the LPA.

2.8.4.1 Guidance on completing the appeal application form

A range of application forms and further guidance is available on the government webpages, as is guidance on how to complete the planning appeal application form (Planning Inspectorate, 2022a).

A valid application for appeal will be decided upon by a Planning Inspectorate inspector. The appeal may be heard by way of written representations, a hearing or an inquiry, or a combination of any of these.

2.8.4.2 After you appeal

The Planning Inspectorate will check the appeal documents for validity and inform you how long it is likely to be before a decision is made (HMG, 2023d).

2.9. Judicial review appeals against the LPA or Planning Inspectorate

If there has been a legal mistake, a decision may be appealed against by way of judicial review in the High Court. Legal advice should be obtained.

Individuals or groups may challenge the decision. To appeal, the person or group must have 'standing', that is, they must have sufficient interest in the matter. However, a broad approach as to what is 'standing' is taken in environmental matters.

Decisions made on called-in applications (i.e. where the Secretary of State exercises their power to direct the LPA to refer an application to them for a decision), and decisions, orders or directions or the costs associated with them under section 78 of the TCPA 1990 regarding the LPA's or the Secretary of State's decisions, as well as a failure to make a decision may all be challenged in the Planning Court of the High Court (HM Courts & Tribunals Service, 2023). However, the Court's permission must be obtained (section 288 of the TCPA 1990).

2.9.1 Grounds of the claim

It is not enough for an applicant to explain that they disagree with the decision.

The basis of the appeal must be that the decision was unlawful because it was one of, or a combination of

- illegal
- procedurally unfair
- unreasonable or irrational
- incompatible with basic human rights.

2.9.2 Time limits

The appeal and all relevant documents must be made both promptly and not later than six weeks after the date on which the grounds to make the appeal arose and was served on the defendant and any interested parties.

If the appeal is not made promptly, the Court may refuse to hear the application even if it is made within the six weeks. Unless the grounds raised are based on the environmental impact assessment regulations, the six-week filing period starts from the date of the decision, not from when a claimant becomes aware of that decision (*Save Britain's Heritage, R (On the Application Of) v City of London Corporation* [2021] EWHC 3561 (Admin)). Clear justification is needed for any extension of time.

2.9.3 Procedure

The defendant and any interested parties have 21 days to provide summary grounds of defence to the claim. The claim and the defence papers are then given to a judge, who decides whether to give permission for the judicial review proceedings to go ahead. It is not uncommon for a judge to refuse permission. However, one can then apply for an oral hearing before a judge to seek permission for the proceedings.

Permission to proceed is granted if the judge decides that the claim is arguable. The claim is then listed for a full substantive hearing before a single judge to determine the claim.

If the Court quashes the decision, the matter is sent back to the Secretary of State to be redetermined. However, the redetermination may not result in the original decision being reversed.

Box 2.2 *Suliman, R (On the Application Of) v Bournemouth, Christchurch and Poole Council* [2022] EWHC 1196 (Admin)

This was a judicial review of the decision of the Bournemouth, Christchurch and Poole Council to grant full planning permission for a mixed-use development, including 130 residential dwellings, on the site of the former police station in Christchurch.

The claimant's property backed onto the site in question. The grounds of her objection to the planning permission were that

1. the council was wrong to proceed on the basis that it had no power to impose a condition requiring that the ecological corridor along the north-west boundary of the site should be 'at least 12 m in width'
2. the council acted in breach of the claimant's legitimate expectation by failing to conduct a visit to her property to review the impact of the proposed development on the outlook towards the site.

The judge rejected the claim on the basis that if the Council had imposed the condition to extend the ecological corridor to 12 m this would be a 'development' that had not been applied for. Therefore, the Council would breach the Wheatcroft principle (*Bernard Wheatcroft Limited v Secretary of State for the Environment* [1982] 43 P&CR 233 at p. 240, *Johnson v Secretary of State for Communities and Local Government* [2007] EWHC 1839 (Admin)).

The judge also rejected the claim on the basis that it was unreasonable 'as it conflicted with the description of the development and the layout plan which the IP was bound to implement if the application for planning permission was granted'.

The judge considered ground 2, and decided that the claimant had failed to establish that the council made a clear, unambiguous and unqualified representation that the committee would visit her property, and so no basis for a legitimate expectation arose.

2.10. Planning controls – breaches and LPA enforcement

If planning permission should have been applied for any building works but it has not been applied for, then there is a planning breach.

There is also a planning breach if planning permission was given subject to conditions and one or more of those conditions has not been complied with.

If planning permission had not been applied for, and building works have started, then the LPA often allows a retrospective application to be made.

2.10.1 Retrospective applications

Section 73A of the TCPA 1990 provides for an application to be made to an LPA for planning permission for a development that has already been carried out. It covers situations where development has been carried out without permission as well as where consented development has been carried out, or where some condition required by a granted planning permission has not been carried out.

It may be considered that the law should be changed to enable the prosecution of owners who proceed with development without first securing planning permission, but such a change in the law has not been made.

Retrospective applications are a mechanism to regularise matters. However, if planning permission is refused or if the owner/applicant declines to make a retrospective application, the LPA may take enforcement action.

Box 2.3 *Malvern Hills District Council v Secretary of State for Housing, Communities and Local Government and Anor* [2021] EWHC 129

On 11 July 2019, Malvern Hills District Council refused planning permission to retain a building erected to store an ex-British Railways steam-operated crane. The building had been erected across a public footpath.

An appeal was made against an enforcement notice to the Planning Inspectorate. The planning inspector granted retrospective planning permission for the erection of a storage shed.

There was then an appeal against this decision to the High Court. The ground of appeal was based on paragraph 98 of the National Planning Policy Framework (NPPF):

> Planning policies and decisions should protect and enhance public rights of way and access, including taking opportunities to provide better facilities for users, for example by adding links to existing rights of way networks including National Trails.

The High Court held that the planning inspector did not fail to treat the need to protect public footpaths as a material consideration, as required under the NPPF. The inspector's grant of planning permission did not authorise the obstruction of the footpath, because the footpath could easily be diverted to an alternative route. Therefore, this ground of appeal was rejected.

2.10.2 Enforcement notices

If planning permission has been applied for but the application has been rejected, or if a retrospective application has been rejected, and work has started, the LPA can issue an enforcement notice instructing that the work must be undone (e.g. that the building must be demolished).

In deciding whether an enforcement notice should be issued, the LPA will consider whether planning control rules have been broken and whether issuing a notice is in the public interest. That is, it will consider whether what has been done is harmful or unacceptable in relation to its effect on public amenities or the existing use of the land in that neighbourhood.

2.10.3 Challenges to enforcement notices

It is illegal not to comply with what the enforcement notice says. However, just as you may appeal the LPA's decision to refuse planning permission, you may appeal against an enforcement notice. An appeal is made under section 289 of the TCPA 1990 to the High Court.

An application for permission to appeal must be made within 28 days of the enforcement decision. However, if an appeal is unsuccessful, it is necessary to comply with the enforcement notice. If there is failure to comply with the notice, there may be a prosecution.

2.11. Nationally significant infrastructure projects (NSIPs)

NSIPs are large-scale developments concerning energy, transport, water or waste in England and Wales. They may involve, for example, developments concerning renewable energy projects, such as windfarms, or airport extensions or nuclear power stations. However, proposals for onshore wind farms of over 50 megawatts were removed from the NSIP regime in 2016 (UK Parliament, House of Commons Library, 2022).

The government agency responsible for examining applications for NSIPs is not the local LPA but the Planning Inspectorate. However, various LPAs may be affected by the NSIP proposal. Each of these will be contacted by the Planning Inspectorate. The LPAs may decide to discuss and coordinate their response with the Inspectorate (Planning Inspectorate, 2022c).

2.11.1 The NSIP planning process

An NSIP goes through a process under the Planning Act 2008 called a development consent order (DCO) process. This process combines the planning permission process discussed above with other types of required consent processes. For example, waste-to-energy plant capable of generating electricity in excess of 50 megawatts are 'nationally significant projects' for the purposes of the Planning Act 2008, and therefore a DCO needs to be obtained, rather than planning permission. The reason is to produce a quicker result than if each consent were applied for separately.

2.11.2 Timescale

The whole process should take around 15 months. There are six stages in the process.

1. **Pre-application**. Potential applicants, such as the Highways Agency, must consult with statutory consultees (as defined by schedule 1 of the Infrastructure Planning (Applications: Prescribed Forms and Procedure) Regulations 2009) and the local community. This is an opportunity for stakeholders to respond and influence the design, layout or location of an NSIP, and any member of the public who makes a 'relevant representation' becomes an 'interested party' at this stage.
2. **Acceptance**. The Planning Inspectorate has 28 days to decide whether or not to accept the application for pre-examination.
3. **Pre-examination**. The public is able to register with the Planning Inspectorate to become an interested party by making a relevant representation, which is a summary of a person's views on an application, made in writing. An examining authority is appointed at this stage. All interested parties will be invited to attend a preliminary meeting, which is run and chaired by the examining authority. There is no statutory timescale for this stage of the process, although it usually takes approximately three months from the applicant's formal notification and publicity of an accepted application.
4. **Examination**. There is no public enquiry, as this is a written process. However, stakeholders, including the public and public consultees, may make written representations to the Planning Inspectorate. The latter has six months to examine an application and three months to make its recommendation to the Secretary of State.
5. **Decision**. The Secretary of State has a further period of three months in which to issue a decision.
6. **Post-decision**. After the Secretary of State has issued a decision, there is only six weeks to challenge it. A challenge is made by way of a judicial review in the High Court.

For more detailed information about any stages of the process, study the Planning Inspectorate webpages (Planning Inspectorate, 2022d, 2022e).

Box 2.4 The Sizewell C project development consent decision

The application involved the construction of a new nuclear power station for producing reliable, low-carbon electricity to help Britain achieve net zero. It is intended that Sizewell C will generate enough low-carbon electricity to supply 6 million homes.

The application was submitted to the Planning Inspectorate for consideration by NNB Nuclear Generation (SZC) Limited on 27 May 2020, and was accepted for examination on 24 June 2020. On 20 July 2022, the Sizewell C project application was granted development consent by the Secretary of State for Business, Energy and Industrial Strategy.

Following an examination during which the local community, statutory consultees, interested parties and the local authorities (East Suffolk District Council and Suffolk County Council) were given the opportunity to give evidence to the examining authority, recommendations were made to the Secretary of State on 25 February 2022.

This is the 114th NSIP and 69th energy application to have been examined by the Planning Inspectorate within the timescales laid down in the Planning Act 2008.

The local authorities – East Suffolk District Council and Suffolk County Council – and other interested parties were able to participate fully.

> The examining authority listened and gave full consideration to local views and the evidence gathered during the examination before making its recommendation.

The decision, the recommendation made by the examining authority to the Secretary of State and the evidence considered by the examining authority in reaching its recommendation are publicly available on the project pages of the National Infrastructure Planning website (National Infrastructure Planning, 2023).

2.12. Environmental permits – pollution control

Construction industry activities may result directly or indirectly in discharges or waste which pollute the land, water or air. There is an integrated Environmental Permit Regime under the Environmental Permitting (England and Wales) Regulations 2016 operating in England and Wales (EPR) to regulate such activities by means of 29 schedules. This section of the chapter focuses on the regime in England regulated by the Environment Agency.

Following the Waste and Environmental Permitting etc. (Legislative Functions and Amendment etc.) (EU Exit) Regulations 2020 amendments to the Environmental Permitting (England and Wales) (Amendment) (EU Exit) Regulations 2019, waste permits and licences are also managed by environmental permitting. Licences and authorisations under the former regimes have been converted to environmental permits under the EPR.

2.12.1 Which activities are regulated

The EPR require that the operator of a regulated facility in England or Wales must have an environmental permit. The regulations concern a wide range of industrial and power-generation activities and 'installations'.

Installations

An installation is:

- a stationary technical unit (STU) where one or more activities listed in part 2 of schedule 1 to the regulations are carried out and
- any other location on the same site where any other directly associated activities (DAA) are carried out which have a technical connection with the activities carried out in the STU and which could have an effect on pollution.

A regulated facility is a site where waste is recycled, stored, treated or disposed of (e.g. a landfill site, a large chicken farm, a food factory, a furniture factory, a dry cleaning shop, a petrol station and a waste operation) and include the main activities given in schedule 1 (i.e. medium combustion plant, small waste incineration plant, mining waste operations, water discharge activities, and groundwater and radioactive substance activities). The emissions and energy efficiency of the facilities are regulated, whereas installations which concern lessor pollutant activities are regulated in terms of air emissions.

Part 2 of schedule 1 to the EPR lists permit-controlled activities that are the basis of an installation. It includes part A activities which concern

- energy – combustion, gasification, liquification and refining activities
- metals – ferrous metals, non-ferrous metals, surface treating metals and plastic materials
- minerals – production of cement and lime, activities involving asbestos, manufacture of glass and glass fibre, other minerals and ceramics
- chemicals – organic, inorganic, fertiliser production, plant health products and biocides, pharmaceutical production, explosives production, manufacturing involving ammonia and storage in bulk
- waste management – incineration and co-incineration of waste, landfills, other forms of disposal of waste, recovery of waste, temporary or underground storage of hazardous waste and treatment of waste water
- other – paper, pulp and board manufacture, carbon, tar and bitumen, coating activities, printing and textile treatments, dyestuffs, timber, rubber, food industries and intensive farming.

Such activities can have a range of environmental impacts.

Emissions to air, land and water

- Water discharge activities are defined in schedule 21 of the EPR. They concern specified activities in and around inland fresh waters, coastal waters and territorial waters that increase the risk of pollution.
- Groundwater activities are defined in schedule 22 of the EPR. They concern discharges that pollute or have the potential to pollute groundwaters.

For example, an environmental permit is needed if the work involves carrying out

- a stand-alone water discharge activity that involves releasing polluting liquids to surface water such as a river or stream
- a stand-alone groundwater activity that involves releasing polluting liquids directly or indirectly to water underground
- flood risk activities defined in schedule 25 of the EPR, which includes various activities carried out on or near main rivers, flood plains, sea defences and remote defences that may affect flooding and drainage.

Energy management and efficiency

- Energy efficiency plays a key role in addressing climate change. It concerns reducing both direct emissions from fossil fuel combustion or consumption, and indirect emissions reductions from electricity generation. See schedule 24 of the EPR.

Waste reduction

- Small waste incineration plant are defined in regulation 2 of the EPR as a waste incineration plant or waste co-incineration plant with a capacity of less than or equal to 10 tonnes per day for hazardous waste or 3 tonnes per hour for non-hazardous waste.

Raw materials consumption

- Combustion plants with a thermal rated input of 1 megawatt or more and less than 50 megawatts ('medium combustion plants'), and
- plants used to generate electricity with a rated thermal input below 50 megawatts ('generators').

Noise and vibration

See Section 3.3 in Chapter 3, and Environment Agency (EA, 2022a).

Accident prevention

- An activity which involves radioactive substances – see schedule 23 of the EPR.

Part B permits control activities that send emissions into the air. Schedule 14 of the EPR applies in relation to every solvent emission activity, but it does not apply to installations used solely for research activities, development activities or the testing of new products or processes.

2.12.1.1 Medium combustion plants and generators and air quality

Following the 2018 amendments to the EPR 2016, permits are now required for the operation of medium combustion plants and certain generators in order to regulate emissions to protect air quality limits on the emissions of certain pollutants into the air from medium combustion plants (the Medium Combustion Plants Directive), and to introduce domestic emission controls in respect of certain types of electricity generator. Other 2018 amendments impose basic safety standards for protection from exposure to ionising radiation.

The government website has information about when you need a permit to operate medium combustion plants and generators to meet air quality requirements (EA, 2022b).

2.12.2 Is a permit needed and who regulates them?

All the above activities and installations are required by law to operate under an environmental permit unless the activity is exempt or excluded from needing one.

Only one permit is needed to cover multiple activities on one project site. The regulators are set out below.

- Part A(1) activities are regulated by the Environment Agency, and part A(2) and part B activities are regulated by local authorities. What constitutes a regulated facility and the classification given in the previous section is explained in *Understanding the Meaning of Regulated Facility* (EA, 2015).
- Natural England is the regulator for activities requiring authorisation under the Conservation of Habitats and Species Regulations 2017 (as amended).
- The Health and Safety Executive is the regulator for UK REACH (HSE, 2023), nuclear installations, offshore oil and gas installations, and sewerage undertakers for trade effluent discharge consents.

2.12.2.1 Who applies for the permit?

Only the 'legal operator' may apply for an environmental permit for the activity (e.g. a water discharge or groundwater). Joint legal operators are allowed.

A legal operator is a legal entity, so it may be an individual, a company limited liability partnership or a government body. However, the operator must have sufficient control of the activity, such as

- has day-to-day control of the activity, including the manner and rate of operation
- ensures that permit conditions are complied with
- decides who holds important staff positions and have incompetent staff removed if required
- makes investment and financial decisions that affect the performance of the activity or how the activity is carried out
- ensures that regulated activities are controlled in an emergency.

While a contractor doing work on the site may be sufficiently in control of the activity to become the legal operator, a remote holding company is unlikely to have sufficient control over it (EA, 2023a).

The death of an operator means the permit may be transferred. However, for the permit to remain valid, a transfer application must be sent to the Environment Agency within six months of the operator's death.

2.12.3 Permit applications and permit exemptions

Free basic pre-application advice is available in England from the Environment Agency (enquiries@environment-agency.gov.uk). Advice includes whether a bespoke or a standard permit is needed, and whether any risk assessments need to be done before submission and the correct application charge. The enquiry should include the location and postcode (or National Grid reference) of the site where the work will be carried out.

The online government guidance on environmental permits (EA, 2022a) provides answers to most of the questions one may have about permits, and has quick links to detailed guidance and online application forms.

2.12.4 Permits and risk assessments and the Environment Agency

The Environment Agency can advise on whether an activity may have an impact on heritage and nature conservation sites or protected species and habitats, and can supply more complex advice such as the monitoring requirements or the requirements for NSIPs. The Agency will provide a cost estimate before it starts work on a request.

2.12.5 Protected sites, habitats and species

If the proposed project is in an area where the development or activity may damage protected sites, habitats or species, the developer or their agent should carry out a risk assessment involving an ecological survey prior to the permit application to find out if there are any protected species that might be affected.

The relevant LPA(s) may not grant planning permission or a permit if the development or activity is likely to damage a protected area or site, species or other wildlife. The Environment Agency can provide a map and information pack to identify if there are any protected areas/sites, species or other wildlife on or near to your proposed activity which are not on the MAGIC map available online (Natural England, 2023). The Agency may also be contacted for advice via its pre-application advice service.

Many species, including birds and animals and their habitats, are protected under the Wildlife and Countryside Act 1981 (as amended), which is the primary legislation in the UK. The Act, when first enacted, implemented the Bern Convention (Council of Europe, 1979) and the European Birds Directive (European Commission, 1979). However, some of the law laid down in the Act differs in Scotland, Wales and Northern Ireland, where additional legislation has been introduced.

The main offences are set out in sections 1, 5 and 8. They include broad protections that prohibit any interference through taking, injuring, killing it or its eggs or offspring, or disturbing places used for shelter and protection. The offence may involve an unlimited fine and/or imprisonment.

2.12.5.1 Sites of Special Scientific Interest (SSSIs)

Under section 28 of the Wildlife and Countryside Act 1981.

> Where Natural England are of the opinion that any area of land is of special interest by reason of any of its flora, fauna, or geological or physiographical features, it shall be the duty of Natural England to notify that fact
>
> (*a*) to the local planning authority (if any) in whose area the land is situated
> (*b*) to every owner and occupier of any of that land, and
> (*c*) to the Secretary of State.

These areas are known as Sites of Special Scientific Interest (SSSIs).

Special areas of conservation (SACs) and special protection areas (SPAs) are also protected areas in the UK. They are designated as such under

- the Conservation of Habitats and Species Regulations 2017 (as amended) in England and Wales (including the adjacent territorial sea) and to a limited extent in Scotland (reserved matters) and Northern Ireland (excepted matters)
- the Conservation (Natural Habitats, &c.) Regulations 1994 (as amended) in Scotland
- the Conservation (Natural Habitats, etc.) Regulations (Northern Ireland) 1995 (as amended) in Northern Ireland
- the Conservation of Offshore Marine Habitats and Species Regulations 2017 in the UK offshore area.

See Joint Nature Conservation Committee (JNCC, 2023) for more information.

2.12.5.2 Other protected areas and sites

The following areas and sites in the UK are also protected.

- National Parks – The National Parks and Access to the Countryside Act 1949 (as amended) provides the framework for the management of National Parks, which are designated by Natural England.
- Areas of Outstanding Natural Beauty (AONB) – part IV of the Countryside and Rights of Way Act 2000 (as amended) (the CRoW Act) provides the framework for the management of AONBs.
- Ramsar wetlands – These are wetlands of international importance for conserving biodiversity. They are recognised by the UK as a signatory of the Ramsar Convention, an intergovernmental treaty that provides the framework for the conservation and wise use of wetlands and their resources (JNCC, 2019).

- Sites in the process of becoming SACs or SPAs ('candidate SACs', 'possible SACs', 'potential SPAs' and sites of community importance (SCIs) or a Ramsar wetland ('proposed Ramsar site').
- Marine Conservation Zones – These are marine nature reserves in UK waters, established under the Marine and Coastal Access Act 2009 (as amended). For proposed development planning, see the government guidance on using marine plans (HMG, 2021).

2.12.5.3 Protected species and biodiversity

The Conservation of Habitats and Species Regulations 2017 (as amended) govern England and Wales (however, see section 2 of the Regulations). Under the regulations, LPAs have a legal obligation to consider the effect on habitats and species when considering whether to allow work to commence on a development. These regulations protect selected species deemed vulnerable to disturbance by building activities including bats, great crested newts, rare reptiles, otters and dormice.

Even though planning permission may be granted, the regulations can require that it is subject to conditions imposed by the LPA. The regulations also set out the responsibilities that local authorities, as well as everyone, have towards European Protected Species (Planning Portal, 2023i).

The Wildlife and Countryside Act 1981 (as amended) protects wildlife (including plants) and controls invasive species.

The Protection of Badgers Act 1992 impacts on groundworks within the vicinity of a sett, even if the works will not require the destruction of a sett. The Act makes it an offence to kill, injure or take a badger, or to damage or interfere with a sett.

The Natural Environment and Rural Communities Act 2006 requires local government to conserve biodiversity and consider it in policy and decision-making. Under section 41 of the Act the Secretary of State is required to publish a list of species and habitats of 'principal importance' for nature conservation.

The Conservation of Habitats and Species Regulations 2017, European Protected Species and Wildlife and Countryside Act 1981 are discussed in Chapter 3, which also covers habitat regulations assessments (Section 3.3.7.1).

Links to comprehensive relevant information about the regulation of the above is available on the government website (HMG, 2023e).

2.12.6 Flood risk and permits

An environmental permission in the form of a permit to carry out flood risk activities is needed if the work is

- in, under, over or near a main river (including where the river is in a culvert), or on or near a flood defence on a main river

- in the flood plain of a main river
- on or near a sea defence.

There are different rules for England, Wales, Northern Ireland and Scotland (HMG, 2023a). Carrying out work without a permit may result in prosecution, prison or a fine.

Any activity that already had a flood defence consent automatically came under environmental permitting rules from 6 April 2016.

For work on or near other watercourses, watercourse consent must be applied for. Application is to the internal drainage board in your area, the Environment Agency or the lead flood authority in your area (HMG, 2023a).

2.12.7 Permit applications and rules

For each activity there is a set of rules and a risk assessment. These are set out on the government webpages for the standard rules for environmental permitting (EA, 2022c).

2.12.7.1 Bespoke permit application

There are different application forms for different activities. For example, there is an application form entitled ‘Part B6 – New bespoke water discharge activity or groundwater activity (point source discharge) or point source emission to water from an installation’, and another called ‘Part B4 – New bespoke waste operation permit’. The various templates and guidance are available on the government webpages (EA, 2023b).

Box 2.5 R (Tarmac Aggregates Ltd) v Secretary of State for Environment, Food and Rural Affairs [2016] Env. L.R. 15

The legal issue in this appeal to the Court of Appeal was whether the proposal to use recovered materials as backfill to reinstate a public footpath at a quarry in Leeds, which restoration was a condition of planning consent, was waste disposal or ‘waste recovery’ under Article 3(15) of the revised Waste Framework Directive. If it was ‘waste recovery’, a standard permit could be issued. However, if it was waste disposal then a bespoke permit would need to be issued. The Court decided that the main purpose of the operation was to ecologically improve the site. Therefore, the Court held that the use of waste materials was waste recovery, and so a standard environmental permit should be issued.

2.12.7.2 Permit exemptions

An exemption allows an operator to carry out a regulated activity without an environmental permit, provided that certain conditions are met. Limits are usually imposed on the scale or type of activity that can take place.

Exemptions must be registered with the regulator. However, there are some non-registrable exemptions, known as ‘non-waste framework directive’ (NWFD) exemptions.

Exemptions cannot be varied or transferred. The regulator also issues regulatory position statements and low-risk waste positions. These are not exemptions.

The government webpages provide detailed guidance about whether there are exemptions for the particular activity so that a permit is not needed. For example, how to classify an effluence and whether an activity is exempt from needing a permit is set out on the government webpage on environmental permits for discharges to surface water and groundwater (EA and Defra, 2022a). However, it is advisable to seek legal advice. It is prudent to advise the Environment Agency of your activity, even if it appears to be an exempt one.

2.12.8 Application form

The type of activity in question must be stated on the permit application. For example, if the activity involves waste water, it must be stated on the application whether the waste water is domestic sewage, trade effluent from a commercial premises, or other polluting matter, and whether it comes from a private residence but is not domestic waste. For example, the discharge from swimming pool drain down or backwash water that is discharged separately from other sources of domestic sewage is not classed as domestic sewage.

2.12.8.1 Management plan

The applicant for a permit must produce a written management plan or management system which is incorporated in the environmental permit. It must identify the risks associated with the applicant's activities, and the steps they will take to minimise those risks.

Operators must apply their management system using the best available techniques or appropriate measures to manage the risks of their activities.

2.12.8.2 Issue of permit

The application, accompanied by the fee and supporting paperwork, is made to the appropriate regulator. This may be the LPA or the Environment Agency.

The determination period begins on the date the regulator receives the application. In many cases, a public consultation will occur before the application is determined.

2.12.8.3 Time limits

The Environment Agency aims to determine all new permit applications within 13 weeks in order to meet a recommendation from the Penfold Review on non-planning consents (BIS, 2010). However, the legal requirements for determining applications are given in paragraph 15 of part 1 of schedule 5 of the EPR. For example

- Two months
  - for an application
  - to transfer a permit
  - for the grant or variation of a permit for a stand-alone flood risk activity.

- Three months for an application for the grant of an environmental permit for
  - mobile plant
  - radioactive substances activities described in schedule 23, part 2, paragraph 11(5)
  - standard facilities (except those that are also part A installations)
  - mining waste operations not involving a mining waste facility to which Article 7 of the Mining Waste Directive applies (schedule 20, paragraph 2(1))
  - certain stand-alone flood risk activities (schedule 5, part 1, paragraph 5(1)).
- Three months for an application to surrender a permit or to vary a permit (other than for a stand-alone flood risk activity or where public participation is required).
- Four months for an application
  - for an environmental permit for any regulated facility except for the exceptions set out above
  - to vary a permit where public participation is required (schedule 5, part 1, paragraphs 5(2) to 5(4)).

2.12.8.4 Issue and conditions

The permit is issued and reviewed by the regulator, which for minor pollutant activities is the local authority, and otherwise the Environment Agency.

If the permit is granted, the regulator can impose such conditions as it sees fit. It can direct the way in which a site must be operated, and the technology and techniques that will be used to minimise the actual and potential impacts on the environment (schedule 5, part 1, paragraph 12(2)).

2.12.9 Environmental permit holding time limits

There is no time limit for holding an environmental permit but it is subject to review by the regulator. A permit can be suspended if a risk of serious pollution arises, and can be revoked when the regulator considers it appropriate.

2.12.9.1 Change, vary or correct, or cancel a permit

It is possible to change (vary) an environmental permit, whether standard or bespoke, or just change your address or contact details. To make changes or to cancel a permit the relevant documents available on the government website (EA and Defra, 2022b) must be completed and sent to the regulator, for example, the Environment Agency.

2.12.9.2 Transfer of a permit

An environmental permit may be wholly transferred to another operator, or partially transferred, as long as the transfer criteria are met.

The application to transfer a permit must be made jointly by the current and future operators and the appropriate fee paid. The fee is not returned if the application is withdrawn.

However, the regulator, usually the Environment Agency, or the LPA for minor activities, must be informed by a joint notification to the regulator and to the transferee where the activity concerns groundwater, flood risk activities or stand-alone water discharge. For all other activities, an application to the regulator must be made jointly by the holder and the proposed transferee.

The regulator may refuse to approve the transfer if it considers that the transferee will not be the operator of the activity, or that the transferee will not comply with the conditions of the permit.

2.12.9.3 Surrender or revocation of a permit

An environmental permit may also be surrendered (i.e. given up). However, surrender procedures must be followed if an operator wishes to surrender an environmental permit. This usually involves carrying out site remediation and monitoring.

There are additional requirements for the surrender of environmental permits relating to landfill sites. The Environment Agency and Natural Resources Wales will only accept an application for surrender when they are satisfied that all necessary measures have been taken to avoid a pollution risk resulting from the operation of the regulated facility, and to return the site to a satisfactory state having regard to the state of the site before the facility was put into operation. In addition, the regulator must be satisfied that at post-completion monitoring the landfill is unlikely to cause a hazard to the environment (EA, 2022c).

Permits can also be revoked by the regulator for non-payment of subsistence fees, or non-compliance. The regulator should be notified of this (EA and Defra, 2021).

2.12.10 Offences and criminal sanctions

All offences occurring on or after 1 January 2017 will be prosecuted under the EPR 2016.

It is an offence to operate without an environmental permit where one is required by law, or to operate without complying with the conditions attached to an environmental permit, or not to comply with an enforcement notice or a suspension notice.

Criminal offences on a summary conviction could involve six months in jail and an unlimited fine. For more serious offences, called 'indictable offences', the penalty may be up to five years in prison and an unlimited fine.

If the offence results in pollution, the regulator can take steps to remedy the pollution but obtain the costs of doing so from the offender. The regulator may also require an enforcement undertaking from the offender or advise of a procedure that must be followed to avoid risk of pollution.

2.12.10.1 Public authority breaches

The new Office for Environmental Protection established under the Environment Act 2021 addresses complaints that a public authority has broken environmental law.

2.12.11 Appeals

An appeal is allowed against the rejection of a permit application, including any conditions imposed, and the revocation of a permit.

An operator can appeal under regulation 31 of the EPR to the Secretary of State against certain decisions made by the relevant regulator. Most appeals are decided by an inspector acting as a delegate of the Secretary of State under section 114(2)(viii) of the Environment Act 1995, to which schedule 20 has effect.

However, the Secretary of State for Environment, Food and Rural Affairs may take over and 'recover' a case if it is particularly important or controversial. If the decision is going to be made by the Secretary of State, the operator will be informed of the reasons why the Secretary of State has decided to 'recover' it.

The appeal must concern an activity or activities covered under the EPR, which include

- an installation (regulation 8(1)(a)) – consists of any 'stationary technical unit' where activities listed in schedule 1 to the regulations and any directly associated activities are carried on
- mobile plant (regulation 8(1)(b))
- a waste operation (regulation 8(1)(c)) – defined as a waste recovery or disposal operation
- a mining waste operation (regulation 8(1)(d)) – the management of extractive waste, whether or not involving a mining waste facility
- a radioactive substances activity (regulation 8(1)(e)) – involving the keeping and use of radioactive material (including mobile radioactive apparatus) or the accumulation and disposal of radioactive waste
- a water discharge activity (regulation 8(1)(f))
- a groundwater activity (regulation 8(1)(g))
- a small waste incineration plant (regulation 8(1)(h)) – all waste incineration plants or co-incineration plants with a capacity less than the thresholds listed in chapter III of the Industrial Emissions Directive and subject to schedule 13 of the EPR 2016
- a solvent emission activity (regulation 8(1)(i)) – an activity listed in annex VII of the Industrial Emissions Directive and subject to schedule 8 of the EPR 2016
- a flood risk activity (regulation 8(1)(j)) – an activity listed in schedule 25 of the EPR 2016.

2.12.11.1 Making the appeal

The appeal form and guidance on the procedure can be obtained from the government website (Planning Inspectorate, 2023) or directly from the relevant regulator.

Under schedule 6(2) of the EPR regulations an appeal requires the following documents

- written notice of appeal
- statement of the grounds of appeal

- statement stating whether you wish the appeal to be dealt with by the written representations procedure or otherwise to be heard by an inspector at a hearing or inquiry
- copy of the relevant application (if any)
- copy of the relevant environmental permit (if any)
- copy of any relevant correspondence, plans etc. that you exchanged with the regulator
- copy of the decision or notice which is the subject of the appeal.

The reasons for the appeal should be given and fully explained on the form, and supported with written evidence (where possible). It should be clearly stated what you want changed about the decision and how it should be changed.

If any of the information enclosed with the appeal has been the subject of a successful application for commercial confidentiality under regulation 48 of the EPR, details should be provided. If this is not done, any documents that have been submitted will be available for public inspection.

Where matters of confidentiality may arise during the life of the appeal, the appellant should disclose this. This case may then be recovered for decision by the Secretary of State.

The notice of appeal should be sent to

The Planning Inspectorate
Environment Appeals Team
3A Eagle Wing
Temple Quay House
2 The Square
Temple Quay
Bristol BS1 6PN
Phone: 0303 444 5584
Email: environment.appeals@planninginspectorate.gov.uk

A copy should be sent to the regulator, which is the Environment Agency for high-risk and part A1 installations (under schedule 7 of the EPR), and to the relevant local authority(ies) for part A2 installations (under schedule 7 of the EPR) and part B installations.

2.12.12 Enforcement

Carrying out regulated activities without the appropriate environmental permit or failure to comply with the permit conditions is illegal. Any offences committed after 1 January 2017 will be prosecuted under the EPR 2016.

The regulator has the power to issue cautions, warnings and, in some cases, fixed penalty notices. However, a prosecution can result in an unlimited fine, imprisonment for individuals, compensation of affected persons, and orders to remove any offending materials. These criminal sanctions are more likely for deliberate or persistent offences.

The EPR set out a range of specific offences and penalties, but it also provides, as an alternative to criminal prosecution, power for the Environment Agency or Natural Resources Wales (or local authority) to take non-criminal steps, including serving (EA and Defra, 2021)

- an enforcement notice
- a suspension notice
- a prohibition notice
- a revocation notice.

References

Statutes

Ancient Monuments and Archaeological Areas Act 1979
Countryside and Rights of Way Act 2000
Environment Act 1995
Environment Act 2021
European Protected Species and Wildlife and Countryside Act 1981
Growth and Infrastructure Act 2013
Local Government Act 2003
Localism Act 2011
Marine and Coastal Access Act 2009 (as amended)
National Parks and Access to the Countryside Act 1949 (as amended)
Planning Act 2008
Planning and Compensation Act 1991
Planning and Compulsory Purchase Act 2004
Planning (Listed Building and Conservation Areas) Act 1990
Planning (Listed Buildings and Conservation Areas) Act 1990
The Natural Environment and Rural Communities Act 2006
The Planning (Wales) Act 2015
The Protection of Badgers Act 1992
The Town and Country Planning Act 1990
Wildlife and Countryside Act 1981

Regulations

Conservation (Natural Habitats, &c.) Regulations 1994 (as amended) in Scotland
Conservation (Natural Habitats, etc.) Regulations (Northern Ireland) 1995 (as amended)
Conservation of Habitats and Species Regulations 2017 (as amended)
Environmental Permitting (England and Wales) Regulations 2016
The Building Regulations 2010
The Conservation of Offshore Marine Habitats and Species Regulations 2017
The Environmental Permitting (England and Wales) (Amendment) (EU Exit) Regulations 2019
The Infrastructure Planning (Applications: Prescribed Forms and Procedure) Regulations 2009
The Town and Country Planning (Control of Advertisements) (England) Regulations 2007
The Town and Country Planning (Tree Preservation) (England) Regulations 2012

Waste and Environmental Permitting etc. (Legislative Functions and Amendment etc.) (EU Exit) Regulations 2020
Town and Country Planning (General Permitted Development) (England) (Amendment) Regulations 2020
Town and Country Planning (Permitted Development and Miscellaneous Amendments) (England) (Coronavirus) Regulations 2020
Town and Country Planning (Use Classes) (Amendment) (England) Regulations 2020

Directives

Industrial Emissions Directive 2010/75/EU
Medium Combustion Plants Directive 2015/2193/EU
The Birds Directive 2009/147/EC
Waste Framework Directive 2008/98/EU

Orders

The Town and Country Planning (Development Management Procedure) (England) Order 2010
The Town and Country Planning (Development Management Procedure) (England) Order 2015
The Town and Country Planning (General Permitted Development etc.) (England) (Amendment) (No. 2) Order 2021
The Town and Country Planning (General Permitted Development) (England) (Amendment) (No. 2) Order 2020
The Town and Country Planning (General Permitted Development) (England) (Amendment) Order 2022
The Town and Country Planning (General Permitted Development) (England) Order 2015
The Town and Country Planning (Use Classes) Order 1987

Case law

Bernard Wheatcroft Limited v Secretary of State for the Environment [1982] 43 P&CR 233
Johnson v Secretary of State for Communities and Local Government [2007] EWHC 1839 (Admin)
Malvern Hills District Council v Secretary of State for Housing, Communities and Local Government and Anor [2021] EWHC 129
R (Tarmac Aggregates Ltd) v Secretary of State for Environment, Food and Rural Affairs [2016] Env. L.R. 15
Save Britain's Heritage, R (On the Application Of) v City of London Corporation [2021] EWHC 3561 (Admin)
Suliman, R (On the Application Of) v Bournemouth, Christchurch and Poole Council [2022] EWHC 1196 (Admin)

Websites

BIS (Department for Business, Innovation and Skills) (2010) GOV.UK (2010) Policy paper. Penfold review of non-planning consents for development: government response. https://www.gov.uk/government/publications/penfold-review-of-non-planning-consents-for-development-government-reponse (accessed 09/03/2023).
Council of Europe (1979) Convention on the Conservation of European Wildlife and Natural Habitats (ETS No. 104). https://www.coe.int/en/web/conventions/full-list?module=treaty-detail&treatynum=104 (accessed 09/03/2023).

DCMS and DSIT (Department for Digital, Culture, Media and Sport and Department for Science, Innovation and Technology) (2022) Code of practice for wireless network development in England. https://www.gov.uk/government/publications/code-of-practice-for-wireless-network-development-in-england (accessed 09/03/2023).

DLUHC (Department for Levelling Up, Housing and Communities) (2014) Tree Preservation Orders and trees in conservation areas. https://www.gov.uk/guidance/tree-preservation-orders-and-trees-in-conservation-areas (accessed 09/03/2023).

DLUHC (2015) Technical housing standards – nationally described space standard. 27 March 2015. https://www.gov.uk/government/publications/technical-housing-standards-nationally-described-space-standard (accessed 09/03/2023).

DLUHC (2019) Advertisements. https://www.gov.uk/guidance/advertisements (accessed 09/03/2023).

DLUHC (2021a) Approved documents. https://www.gov.uk/government/collections/approved-documents (accessed 09/03/2023).

DLUHC (2021b) Planning application and fire statement forms: templates. https://www.gov.uk/government/publications/planning-application-forms-templates-for-local-planning-authorities (accessed 09/03/2023).

DLUHC (2022a) Consultation and pre-decision matters. https://www.gov.uk/guidance/consultation-and-pre-decision-matters#covid19 (accessed 09/03/2023).

DLUHC (2022b) Flood risk and coastal change. https://www.gov.uk/guidance/flood-risk-and-coastal-change#planning-and-flood-risk (accessed 18/08/2022).

EA (Environment Agency) (2015) *Understanding the Meaning of Regulated Facility*. Regulatory Guidance Series No. RGN 2. https://assets.publishing.service.gov.uk/government/uploads/system/uploads/attachment_data/file/964485/LIT_6528.pdf (accessed 09/03/2023).

EA (2022a) Noise and vibration management: environmental permits. https://www.gov.uk/government/publications/noise-and-vibration-management-environmental-permits/noise-and-vibration-management-environmental-permits (accessed 09/03/2023).

EA (2022b) Medium combustion plant: when you need a permit. Environment Agency, Natural Resources Wales, Department for Environment, Food and Rural Affairs, and Welsh Government. https://www.gov.uk/guidance/medium-combustion-plant-when-you-need-a-permit (accessed 09/03/2023).

EA (2022c) Standard rules: environmental permitting. https://www.gov.uk/government/collections/standard-rules-environmental-permitting (accessed 09/03/2023).

EA (2023a) Environmental permits: detailed information. Environment Agency, Department for Environment, Food and Rural Affairs, Offshore Petroleum Regulator for Environment and Decommissioning, Natural Resources Wales, Welsh Government, and others. https://www.gov.uk/topic/environmental-management/environmental-permits (accessed 09/03/2023).

EA (2023b) Environmental permit application form: new bespoke permit. https://www.gov.uk/government/collections/environmental-permit-application-forms-for-a-new-bespoke-permit (accessed 09/03/2023).

EA and Defra (Environment Agency and Department for Environment, Food and Rural Affairs) (2021) How you'll be regulated: environmental permits. https://www.gov.uk/guidance/how-youll-be-regulated-environmental-permits (accessed 09/03/2023).

EA and Defra (2022a) Discharges to surface water and groundwater: environmental permits. https://www.gov.uk/guidance/discharges-to-surface-water-and-groundwater-environmental-permits#check-if-the-waste-water-is-domestic-sewage-or-trade-effluent (accessed 09/03/2023).

EA and Defra (2022b) Change, transfer or cancel your environmental permit. https://www.gov.uk/guidance/change-transfer-or-cancel-your-environmental-permit#cancel-surrender-your-permit (accessed 09/03/2023).

European Commission (1979) Directive 2009/147/EC of the European Parliament and of the Council of 30 November 2009 on the conservation of wild birds. https://ec.europa.eu/environment/nature/legislation/birdsdirective/index_en.htm (accessed 09/03/2023).

HCLG (Ministry of Housing, Communities and Local Government) (2019) The Future Homes Standard: changes to Part L and Part F of the Building Regulations for new dwellings. https://www.gov.uk/government/consultations/the-future-homes-standard-changes-to-part-l-and-part-f-of-the-building-regulations-for-new-dwellings (accessed 09/03/2023).

HCLG (2021) National Planning Policy Framework. https://www.gov.uk/government/publications/national-planning-policy-framework–2 (accessed 09/03/2023).

HSE (Health and Safety Executive (2023) UK REACH explained. https://www.hse.gov.uk/reach/about.htm (accessed 09/03/2023).

Historic England (2023a) Listed building consent. https://historicengland.org.uk/advice/planning/consents/lbc (accessed 09/03/2023).

Historic England (2023b) Scheduled monument consent. https://historicengland.org.uk/advice/planning/consents/smc (accessed 09/03/2023).

HM Courts & Tribunals Service (2023) Planning Court. https://www.gov.uk/courts-tribunals/planning-court (accessed 09/03/2023).

HMG (His Majesty's Government) (2021) Using marine plans. https://www.gov.uk/government/publications/using-marine-plans (accessed 09/03/2023).

HMG (2023a) Check if you need permission to do work on a river, flood defence or sea defence. https://www.gov.uk/permission-work-on-river-flood-sea-defence (accessed 09/03/2023).

HMG (2023b) Get flood risk information for planning in England. https://flood-map-for-planning.service.gov.uk (accessed 09/03/2023).

HMG (2023c) Scheduled monument consent (England, Scotland and Wales). https://www.gov.uk/scheduled-monument-consent-england-scotland-wales (accessed 09/03/2023).

HMG (2023d) Appeal a planning decision. https://www.gov.uk/appeal-planning-decision/after-you-appeal (accessed 09/03/2023).

HMG (2023e) Environmental management: detailed information. https://www.gov.uk/topic/environmental-management (accessed 09/03/2023).

HMG (2023f) Appeal a decision about a lawful development certificate. https://www.gov.uk/appeal-lawful-development-certificate-decision (accessed 09/03/2023).

HMG (2023g) Check the long term flood risk for an area in England. https://www.gov.uk/check-long-term-flood-risk (accessed 09/03/2023).

Natural England, Defra and EA (Department for Food, Agriculture and Rural Affairs and Environment Agency) (2023) Wildlife and habitat conservation: detailed information. https://www.gov.uk/topic/environmental-management/wildlife-habitat-conservation (accessed 09/03/2023).

JNCC (Joint Nature Conservation Committee) (2019) Ramsar Convention. https://jncc.gov.uk/our-work/ramsar-convention (accessed 09/03/2023).

JNCC (2023) Special areas of conservation. https://sac.jncc.gov.uk (accessed 09/03/2023).

MHGLC (Ministry of Housing, Government and Local Communities) (2020) Memorandum to the Town and Country Planning (Permitted Development and Miscellaneous Amendments) (England) (Coronavirus) Regulations 2020. https://www.legislation.gov.uk/uksi/2020/632/pdfs/uksiem_20200632_en.pdf (accessed 09/03/2023).

National Infrastructure Planning (2023) Sizewell C Project. https://infrastructure.planninginspectorate.gov.uk/projects/eastern/the-sizewell-c-project (accessed 09/03/2023).

Natural England (2023) MAGIC. Interactive mapping at your fingertips. https://magic.defra.gov.uk (accessed 09/03/2023).

Planning Inspectorate (2022a) Planning appeals: how to complete your appeal form. https://www.gov.uk/government/publications/planning-appeals-how-to-complete-your-appeal-form (accessed 09/03/2023).

Planning Inspectorate (2022b) *Procedural Guide: Planning Appeals – England*. https://www.gov.uk/government/publications/planning-appeals-procedural-guide/procedural-guide-planning-appeals-england (accessed 09/03/2023).

Planning Inspectorate (2022c) National Infrastructure Planning. https://infrastructure.planninginspectorate.gov.uk (accessed 09/03/2023).

Planning Inspectorate (2022d) National infrastructure planning. The process. https://infrastructure.planninginspectorate.gov.uk/application-process/the-process (accessed 09/03/2023).

Planning Inspectorate (2022e) National infrastructure planning. Advice notes. https://infrastructure.planninginspectorate.gov.uk/legislation-and-advice/advice-notes (accessed 09/03/2023).

Planning Inspectorate (2023) Environmental permit – Guidance on the appeal procedure. https://www.gov.uk/government/publications/environmental-permit-appeal-form/environmental-permit-guidance-on-the-appeal-procedure (accessed 09/03/2023).

Planning Portal (2023a) Use classes. https://www.planningportal.co.uk/info/200130/common_projects/9/change_of_use (accessed 09/03/2023).

Planning Portal (2023b) Change of use. Planning permission. https://www.planningportal.co.uk/permission/common-projects/change-of-use/planning-permission (accessed 09/03/2023).

Planning Portal (2023c) Fee calculator (England). Calculate your planning fee. https://1app.planningportal.co.uk/FeeCalculator/Standalone (accessed 09/03/2023).

Planning Portal (2023d) Listed building consent. https://www.planningportal.co.uk/planning/planning-applications/consent-types/listed-building-consent (accessed 09/03/2023).

Planning Portal (2023e) Scheduled monument consent. https://historicengland.org.uk/advice/planning/consents/smc (accessed 09/03/2023).

Planning Portal (2023f) Advertisement consent. https://www.planningportal.co.uk/planning/planning-applications/consent-types/advertisement-consent (accessed 09/03/2023).

Planning Portal (2023g) Hazardous substances consent. https://www.planningportal.co.uk/permission/responsibilities/other-permissions-you-may-require/hazardous-substances-consent (accessed 09/03/2023).

Planning Portal (2023h) Find and download paper forms. https://www.planningportal.co.uk/info/200126/applications/61/paper_forms (accessed 09/03/2023).

Planning Portal (2023i) European Protected Species. https://www.planningportal.co.uk/permission/responsibilities/other-permissions-you-may-require/european-protected-species (accessed 09/03/2023).

UK Parliament, House of Commons Library (2022) Planning for Nationally Significant Infrastructure Projects. https://commonslibrary.parliament.uk/research-briefings/sn06881 (accessed 09/03/2023).

Welsh Government (2022) National planning policy. Guidance and services. https://gov.wales/national-planning-policy (accessed 09/03/2023).

Welsh Government (2023) National Development Framework, setting the direction for development in Wales to 2040. https://www.gov.wales/future-wales-national-plan-2040 (accessed 09/03/2023).

Baker F and Charlson J
ISBN 978-0-7277-6645-8
https://doi.org/10.1680/elsc.66458.077

Chapter 3
Environmental impact and habitat assessments

Francine Baker

3.1. Introduction

Planning decisions regarding the carrying out of development projects can have a significant impact on the environment. It used to be thought that the environment could adapt and absorb the changes brought about by developments. Now we are aware that a development affects the appearance and quality of the environment. A new development may have long-term adverse effects on the environment. It may bring changes to ecosystems. Whether we are talking about infrastructure projects, the creation of motorways or housing estates, or householder activities such as creating garages or extensions to homes, developments can produce many unforeseen and unexpected impacts on the environment.

3.1.1 Sources of environmental problems from development

The major sources of environmental problems from development include

- those arising from new development with regard to the choice of location, siting, design and demand for transport
- transport
- tourism (its positive contribution to local economies needs to be offset against its effect on the landscape and the fabric of historic buildings, and the increased transport demand)
- agriculture, which uses pesticides and has an effect on habitats and biodiversity
- contaminated land from the legacy of industrial land and brownfield land (see Chapter 4)
- industry, which involves energy consumption and creates pollution
- noise and water pollution (see Section 3.3.7.3 for noise and Chapter 6 for water pollution).

Together, environmental law and planning law and policy seek to address the tension between development and protection of the environment. However, it may seem that in recent times the law has been playing catch up with climate-change events.

3.1.2 Chapter contents summary

This chapter will discuss the main measures taken to address negative impacts on the environment before the development may start. These include a design and access statement (DAS), the standard environmental impact assessment (EIA) report required by planning for certain developments, and the habitat and species, noise and flood risk assessments, which may be required, as well as wildlife licences. It will outline the type of projects that are likely to need an EIA. These assessments also relate to various types of planning permission, and so this chapter should be read in conjunction with Chapter 2 – Planning and environmental permits.

In addition, this chapter will discuss the Environment Act 2021 and its prerequisite requirements before planning permission can be granted. This includes showing a biodiversity net gain, which may be required to be accompanied by covenants.

3.2. When is a design and access statement required?

The legal requirement for a design and access statement (DAS) under the Town and Country Planning (Development Management Procedure) (England) Order 2015 (as amended) (the DMPO) identifies that the design of a proposed development may impact on the environment. The DMPO requires that a local planning authority (LPA) is not permitted by law to consider a planning application if it has not been accompanied by a DAS, where such a statement is required by law.

Article 9(1) of the DMPO requires that a DAS must accompany an application for both outline and full planning application permissions for (a) a major development or (b) a development that concerns a conservation area or a World Heritage Site and involves (i) the provision of one or more dwelling-houses or (ii) the provision of a building(s) for which the proposed floorspace is 100 square metres or more.

Article 9(2) requires that the DAS should explain (a) the design principles and concepts that have been applied to the development, and (b) how issues relating to access to the development have been dealt with.

No DAS is required for applications concerning mining and engineering operations, a change of use of a development or a householder development, except in conservation areas, or where conditions attached to a planning permission have been altered (section 73 of the Town and Country Planning Act 1990), and other developments set out in Article 4C of the Town and Country Planning (General Development Procedure) (Amendment) (England) Order 2010.

3.2.1 The content of a DAS

Under Article 9(3)

> A design and access statement must
>
> (*a*) explain the design principles and concepts that have been applied to the development
> (*b*) demonstrate the steps taken to appraise the context of the development and how the design of the development takes that context into account
> (*c*) explain the policy adopted as to access, and how policies relating to access in relevant local development documents have been taken into account
> (*d*) state what, if any, consultation has been undertaken on issues relating to access to the development and what account has been taken of the outcome of any such consultation, and

(*e*) explain how any specific issues which might affect access to the development have been addressed.

3.2.1.1 LPA involvement

It is more efficient and advisable to obtain advice from the relevant LPA at the planning pre-application stage, rather than later, as the LPA can offer a design review service and guidance on what should be included in the DAS, and also consider if an environment impact assessment is needed (DLUHC, 2019a).

3.3. Environmental impact assessment (EIA)

A more comprehensive effective means of checking that developments do not contribute to damaging our environment, or at least that any damage is mitigated in conjunction with consideration of any social or economic impacts of the development, is through an EIA for a development.

The object is to prevent environmental problems before they arise by requiring an assessment of the environmental effects of certain development projects, resulting in a report, known as an 'environmental statement', that is presented to the relevant LPA before submitting, or along with, the planning application.

3.3.1 When is an EIA required?

The requirement to assess the environmental impact of certain projects was first introduced into the UK in 1985 by a European Community Directive. The EU's Environmental Impact Assessment Directive (2011/92/EU as amended by 2014/52/EU) requires major building or development projects to be assessed for their impact on the environment before the project can start.

In England and Wales, an EIA may be required as part of the planning permission process under the Town and Country Planning (Environmental Impact Assessment) Regulations 2017 and Town and Country Planning (Environmental Impact Assessment) (Wales) Regulations 2017 (as amended) (the TCPEIA), or under the Infrastructure Planning (Environmental Impact Assessment) Regulations 2017 (as amended) (the IPEIA).

In addition, other specific legislation concerning highways, power stations, water resources, marine developments, pipelines, land drainage, and forestry relating to EIAs may need to be complied with. For example, an EIA for a marine development must comply with

- the Marine Works (Environmental Impact Assessment) Regulations 2007 (as amended)
- the Electricity Works (Environmental Impact Assessment) (England and Wales) Regulations 2000 (as amended)
- the Water Resources (Environmental Impact Assessment) (England and Wales) Regulations 2003 (as amended).

Developers, or those appointed by them for any development, should engage early (at the pre-application phase) with statutory consultees and local authorities in order to help address any issues and opportunities to avoid delays occurring at the formal application stage.

3.3.2 Who will decide if an EIA is required?

Usually, the LPA will decide if an EIA is needed. However if the development proposals concern a nationally significant infrastructure project (NSIP) (see Chapter 2) an EIA and any planning permissions will be dealt with by the Secretary of State through the Planning Inspectorate, and not by the LPA.

3.3.2.1 Timeframe for the determination of a planning application with an environmental statement

Where both the valid planning application involving an EIA, and therefore an environmental statement (ES), have been received by the LPA, the application is usually decided within 16 weeks from the date of the following day it was received (regulation 68 of the TCPEIA). The timeframe may be extended by agreement.

However, the planning application will not be considered until the ES has been delivered. Where the applicant has submitted the planning application, but not the ES, the applicant must, before submitting the ES, comply with the publicity requirements of paragraphs (2) to (5) of regulation 20 of the TCPEIA.

3.3.3 Request for a screening opinion

The applicant can request an informal opinion from the planning authority as to whether an EIA is needed for their proposal. The TCPEIA regulations require the planning authority to give a formal opinion as to whether an EIA is required when there has been a request. This is called a 'screening opinion'.

If the applicant is dissatisfied with the screening opinion or if no opinion has been given when requested, or not given within the relevant time, a request may be made to the Secretary of State to provide a screening direction (regulation 5(6) of the TCPEIA). This direction will say whether or not an EIA is required.

3.3.4 Is an EIA required for schedule 1 developments?

An EIA is always needed for those developments, or certain changes or extensions to them, set out in schedule 1 of the TCPEIA and IPEIA regulations.

An ES must then be submitted by the applicant at the time of submitting the planning application. If the ES is not submitted, the planning application will be deemed to be invalid.

Schedule 1 of the IPEIA includes, among other things, proposals for the development of

- Crude-oil refineries (excluding undertakings manufacturing only lubricants from crude oil) and installations for the gasification and liquefaction of 500 tonnes or more of coal or bituminous shale per day. More recently, there are the Offshore Oil and Gas Exploration, Production, Unloading and Storage (Environmental Impact Assessment) Regulations 2020.
- Thermal power stations and other combustion installations with a heat output of 300 megawatts or more.
- Nuclear power stations and other nuclear reactors (except research installations for the production and conversion of fissionable and fertile materials, whose maximum power does not exceed 1 kilowatt continuous thermal load).

- Installations for the reprocessing of irradiated nuclear fuel.
- Integrated works for the initial smelting of cast iron and steel.
- Installations or the production of non-ferrous crude metals from ore, concentrates or secondary raw materials by metallurgical, chemical or electrolytic processes.
- Installations for the extraction of asbestos and for the processing and transformation of asbestos and products containing asbestos.
- Waste disposal by incineration.
- Major wastewater treatment works.
- Pipelines with a diameter of more than 800 millimetres and a length of more than 40 kilometres for the transport of (a) gas, oil or chemicals, or (b) carbon dioxide streams for the purposes of geological storage.
- Works for the transfer of water resources, other than piped drinking water.
- Industrial plant for (a) the production of pulp from timber or similar fibrous materials, or (b) the production of paper and board with a production capacity exceeding 200 tonnes per day.
- Quarries and open-cast mining where the surface of the site exceeds 25 hectares, or peat extraction where the surface of the site exceeds 150 hectares.
- Construction of overhead electrical power lines with a voltage of 220 kilovolts or more and a length of more than 15 kilometres.
- Chemical works.
- Power stations.
- Railways.
- Airports.
- Major roads.

Development proposals where the sole purpose concerns military or national defence or in response to a civil emergency do not usually require an EIA. However, see regulations 60–63 of the TCPEIA for more details.

3.3.5 Is an EIA required for schedule 2 developments?

The TCPEIA and IPEIA regulations set out the types of development proposals that are subject to a screening test to determine if an EIA is required. However, an applicant may request that the proposal is subject to an EIA even if one is not required.

Schedule 2 of the TCPEIA sets out the descriptions of development and the thresholds and criteria for defining what is a schedule 2 development. An EIA for a schedule 2 development will be required if the LPA or Secretary of State considers that the project would be likely to give rise to significant effects on the environment by virtue of its nature, size or location.

There is a wide range of schedule 2 developments (HMG, 2017). Many have a size or other type of threshold. The developments include, among others

- Urban developments, including holiday villages, shopping centres, car parks, sports stadiums, leisure centres and multiplex cinemas, where (i) the development includes more than 1 hectare of urban development which is not dwelling-house development,

(ii) the development includes more than 150 dwellings, or (iii) the overall area of the development exceeds 5 hectares.

- Railways (unless included in schedule 1), where the area of the works exceeds 1 hectare
- Airfields (unless included in schedule 1), where (i) the development involves an extension to a runway, or (ii) the area of the works exceeds 1 hectare.
- Large scale or groups of wind turbines.
- Rubber production.
- Chemical production.
- Wastewater treatment plants (unless included in schedule 1), where the area of the development exceeds 1000 square metres.
- Quarries and opencast mines.
- Coke ovens.
- Foundries and forges.
- Surface industrial installations, including hydraulic fracking.

A summary diagram setting out all the schedule 2 criteria and thresholds against which to assess whether your proposal is subject to a screening process is available (HMG, 2021a). It may be helpful as a quick reference, but always consult with your LPA to check if an EIA is required.

3.3.5.1 Three-stage process to assess if an EIA is needed for a schedule 2 development

Planning authorities will consider

1. Is the type of development listed in a schedule 2 category?
2. If it is, does it exceed the threshold given for that category? For example, for industrial installations for carrying gas, steam and hot water, the threshold given in schedule 2 is that the area of the works exceeds 1 hectare. Otherwise, consider whether the development is in a sensitive area such as an Area of Outstanding Natural Beauty, or a special area of conservation (SAC) or a Site of Special Scientific Interest (SSSI).
3. If the answer to 2 is yes, is the development likely to have significant effects on the environment with regard to the factors set out in schedule 3?

3.3.5.2 Schedule 3 criteria applied to a schedule 2 development

In accordance with regulation 5(4) of the TCPEIA, schedule 3 of this Act sets out the selection criteria that government authorities (LPA or Planning Inspectorate) use to screen a schedule 2 development to decide if it requires an EIA. This part of the TCPEIA regulations is set out below.

Schedule 3

1. The characteristics of development must be considered with particular regard to

   (*a*) the size and design of the whole development
   (*b*) cumulation with other existing development and/or approved development
   (*c*) the use of natural resources, in particular land, soil, water and biodiversity

(*d*) the production of waste
(*e*) pollution and nuisances
(*f*) the risk of major accidents and/or disasters relevant to the development concerned, including those caused by climate change, in accordance with scientific knowledge
(*g*) the risks to human health (for example, due to water contamination or air pollution).

2. (1) The environmental sensitivity of geographical areas likely to be affected by development must be considered, with particular regard, to

(*a*) the existing and approved land use
(*b*) the relative abundance, availability, quality and regenerative capacity of natural resources (including soil, land, water and biodiversity) in the area and its underground
(*c*) the absorption capacity of the natural environment, paying particular attention to the following areas

(i) wetlands, riparian areas, river mouths
(ii) coastal zones and the marine environment
(iii) mountain and forest areas
(iv) nature reserves and parks
(v) European sites (conservation areas) and other areas classified or protected under national legislation
(vi) areas in which there has already been a failure to meet the environmental quality standards, and relevant to the project, or in which it is considered that there is such a failure
(vii) densely populated areas
(viii) landscapes and sites of historical, cultural or archaeological significance.

The answer to stage 3 of the process in Section 3.3.5.1 will be a matter of opinion. Different authorities may reach different opinions as to whether an EIA is needed for a schedule 2 development because of different interpretations of the schedule 3 factors.

Box 3.1 *Berkley v Secretary of State for the Environment and others* [2000] UKHL 36

The House of Lords ruled that an EIA required a separate process, which must be applied. Planning authorities therefore can no longer effectively argue that, although they had not required an EIA, their decision not to require one is an EIA determination. A court is therefore not entitled retrospectively to dispense with the requirement of an EIA on the ground that the outcome would have been the same.

This approach has been confirmed in later case law, such as *R (On the Application Of Champion) v North Norfolk District Council & Anor* [2015] UKSC 52 at [51, and 53]. There, Lord Carnwath delivering the Court's judgement, stated that the initial unresolved uncertainties over the mitigation measures mandated an EIA. Although the measures were ultimately resolved to the satisfaction of Natural England and others, this did not mean that there had been no need for an EIA. There should have been an EIA, and this was a procedural irregularity that was not cured by the final decision.

3.3.6 The process if an EIA is required for a development

The legislative process of an EIA (regulation 4 of the of the TCPEIA and regulation 5 of the IPEIA) includes a prediction about the effects and likely effects of the development on the environment, and identifying measures to be taken to reduce or avoid any adverse effects.

Where a development falls under schedule 1, or if the planning authority decides that a schedule 2 developments requires an EIA, then the legislative process consists of

(*a*) the preparation of an ES by the applicant (see Section 3.4)
(*b*) any consultation, publication and notification required by, or by virtue of, the IPEIA regulations or any other enactment in respect of an EIA development (see Section 3.4.5.1)
(*c*) the steps required by the regulations (see Sections 3.4 and 3.5).

3.3.7 The EIA process and the Conservation of Habitats and Species Regulations 2017

It is advisable to consult with the relevant LPA(s) and the relevant statutory nature conservation body (i.e. Natural England for projects in England) from the conception of the project, and at each stage of the development's progress.

The Conservation of Habitats and Species Regulations 2017 consolidate and update the Conservation of Habitats and Species Regulations 2010. The 2017 regulations transposed the European Directive 92/43/EEC (the Habitats Directive) on the conservation of natural habitats and of wild fauna and flora and elements of Directive 2009/147/EC (the Birds Directive) on the conservation of wild birds in England, Wales and, to a limited extent, Scotland and Northern Ireland. The objective of the Habitats Directive is to protect biodiversity through the conservation of natural habitats and species of wild fauna and flora. It lays down rules for the protection, management and exploitation of such habitats and species (HMG, 2017).

Some changes were made by the Conservation of Habitats and Species (Amendment) (EU Exit) Regulations 2019 to the Conservation of Habitats and Species Regulations 2017 (as amended) (the CHSR). Most of them concern the transference of functions and operation from the European Commission to England and Wales. They establish management objectives (called the 'network objectives') for the national site network. These objectives contribute to the conservation of UK habitats and species as well as those that are of pan-European importance (CIEEM, 2021).

The CHSR is designed to protect against LPA plans and project proposals that could significantly damage certain features of many sites in the UK and the offshore. They include, among others, special areas of conservation (SACs), which include certain woods, fens and estuaries (JNCC, 2021a), and special areas of protection (SAPs) for birds, which include certain moors, forests, firths, lochs, loughs and estuaries, coasts and islands (JNCC, 2021b) and wetlands of international importance designated under the Ramsar Convention (JNCC, 2019), as well as other areas which concern compensation for sites damaged.

3.3.7.1 Habitat regulations assessment (HRA)

Any development proposal (including a plan or a project) affecting current, proposed or potential SACs or SAPs or a Ramsar site usually requires an HRA. The proposal may affect one of the protected sites if the proposed development is on the site, near the site or even some distance away, as it may be likely to cause water, air or noise pollution (Defra and Natural Resources Wales, 2021) You can check if there is an impact risk zone (IRZ) near or around your project site on the MAGIC map (Natural England, 2023).

The term 'proposal' is given a wide interpretation under the CHSR. For example, it applies to

- applications for planning permission and to housing, retail or industrial projects, or to changes to a project, building or installing transport schemes, the extraction of timber, minerals or water, and wind farms
- activities under a permitted planning permission, and for licences, permits and consents issues under council byelaws
- the maintenance of highways and flood defences, and repairing underground cables and keeping powerlines clear.

It is the responsibility of a 'competent authority' to carry out an assessment under the CHSR if a development proposal concerns a plan or a project. Such authorities include an LPA, statutory authorities such as a water or energy company, a planning inspector, a government minister or government department, or an ombudsman.

Impact risk zones can be checked to see if your proposal may affect one of the protected sites by looking at the MAGIC map (Natural England, 2023).

The CHSR also protect European protected species (EPS), which include

- all species of bats
- beavers
- great crested newts
- hazel or common dormice
- otters
- natterjack toads
- reptiles (some species)
- protected plants (some species)
- large blue butterfly
- sturgeon.

The Wildlife and Countryside Act 1981 (as amended) (the WCA) protects an extensive range of wild animals, as set out in schedules 5 and 6 of the Act.

In England, EPS are protected by criminal offences under both the CHSR and the WCA.

Under regulation 43(1) of the CHSR, the key EPS offences are

(*a*) deliberately capture, injure or kill any wild animal of a European protected species
(*b*) deliberately disturb wild animals of any such species
(*c*) deliberately take or destroy the eggs of such an animal, or
(*d*) damage or destroy a breeding site or resting place of such an animal.

Under section 9(4)(b) of the WCA a person is guilty of an offence if they intentionally or recklessly disturb an EPS while it is occupying a structure or place that it uses for shelter or protection.

Under section 9(4)(c) of the WCA a person is guilty of an offence if they intentionally or recklessly obstruct access to any structure or place that any EPS uses for shelter or protection.

If a development activity is likely to trigger an offence under regulation 43 of the CHSR, an EPS licence is required.

The changes under section 10 of the WCA from 30 September 2022 mean that, in England, there is no exposure to the offences under sections 9(4)(b) and 9(4)(c) of the WCA if a Conservation of Habitats and Species Regulations 2017 EPS licence is obtained, followed and complied with. The penalty for non-compliance is an unlimited fine and up to six months in prison for each offence, as well as the costs of prosecution.

3.3.7.1.1 Wildlife licences for development works. Wildlife licences may be given under the above legislation for what would otherwise be prohibited activity. Applications for licences can be made once full planning permission has been granted, as long as any planning conditions or reserved matters relating to wildlife have been fulfilled. It is advisable to engage a professional ecological consultant to conduct the surveys to provide information for the licence application.

Section 111 of the Environment Act 2021 amended the wildlife licensing regime in England given under section 16 of the Wildlife and Countryside Act 1981, and the amendment came into force from 30 September 2022. This means that that for development works which may interfere with animals and plants domestically protected under the WCA (e.g. water voles) a licence may be obtained under the WCA for the statutory purpose of 'reasons of overriding public interest'.

Consequently, a licence for development purposes may be applied for under the new statutory licence purpose of 'reasons of overriding public interest', whereas previously it was not possible to obtain a licence for development works under the WCA if, for example, the works may have disturbed a species, even though not deliberately, as there could still be a disturbance offence under section 9(4)(b) of the WCA.

Therefore, there are now three tests that must be met to obtain a licence

- 'reasons of overriding public interest' (the purpose) test
- no 'satisfactory' alternative test
- favourable conservation status test.

However, it must be shown that the effects of the environmental impact will be mitigated to ensure the maintenance of the population of the species under consideration.

There is a range of different wildlife licences, and which are required depends on the work to be done and its potential impact on a species. These include, but are not limited to, a general licence for low-risk work, a class licence when there is a need for expertise concerning the proposed work, an individual licence for proposed work that is not covered under a general or class licence, and an organisational licence. For more details and the application forms for licences, see the government webpages (Natural England and Defra, 2022). The application form for a mitigation licence regarding an EPS can also be obtained from the government website (Natural England, 2022).

Licences should be applied for to the relevant statutory conservation body. In England this is Natural England, in Wales it is the Natural Resources Body for Wales, and in Scotland it is Scottish Natural Heritage.

A licence decision may be issued within 30 days of application and there may be a fee attached.

From 30 September 2022, the period of validity of a wildlife licence in England (issued by Natural England) has been extended to five years.

3.3.7.1.2 Enforcement of licences. If an activity is carried out without a wildlife licence there is a penalty of an unlimited fine and up to six months in prison.

Pre-submission advice and screening is available from the Natural England licensing team (Natural England, 2019).

3.3.7.2 Coordination between an EIA and an HRA

A development may be subject to an EIA, as well as an HRA (under regulation 63 of the CHSR).

The LPA or the Secretary of State, where appropriate, must ensure that the HRA and the EIA are coordinated (regulation 27 of the TCPEIA).

Box 3.2 *R (On the Application Of Champion) v North Norfolk District Council & Anor* [2015] UKSC 52

One of the issues in this case concerned the timing of the decision as to whether an assessment under the European Habitats Directive (Directive 92/43/EEC) or the Environmental Impact Assessment Directive was required.

On this particular issue, Lord Carnwath, delivering the court's judgement, reasoned that the formal procedures prescribed for EIA purposes in the TCPEIA regulations were not replicated in the habitats legislation. There was, therefore, nothing in the language of the Habitats Directive or in the decisions of the Court of Justice of the European Union (CJEU) to support the view that a separate 'screening' stage was required. Lord Carnwath stated that all that is required is that there should be an appropriate assessment where there is found to be a risk of significant adverse effects to be protected. No special procedure was required, but a high standard of investigation was necessary. He said that the decision regarding the issue ultimately rested with the authority's judgement. As the planning authority and the expert consultees were satisfied that the material risk of significant effects on the river had been eliminated, there was no reason to think that the conclusion would be different if they had decided from the outset that appropriate assessment was required.

Box 3.3 *Hudson, R (On the Application Of) v Royal Borough of Windsor and Maidenhead & Ors* [2021] EWCA Civ 592

The fifth ground of appeal from the court of first instance concerned whether a different decision might have been reached had the council carried out an 'appropriate assessment' under the Habitats Directive and the Habitats Regulations 2017. It was argued by Mr Hudson, former chairman of the Berkshire branch of the Campaign to Protect Rural England, that permission for the project to go ahead should not have been given to Legoland Windsor Resort and to Merlin Attractions Operations and Merlin Entertainments by the Royal Borough of Windsor and Maidenhead.

Coulsen LJ considered that it would be a 'very unusual case in practice in which the court's consideration of the seriousness of the breach and any prejudice caused was sufficient to establish that it was highly likely that the outcome would not have been substantially different, but insufficient to establish that the result would have been the same' (101). Applying those principles, Coulson LJ found that the legal burden under section 31(2A) of the regulations had been discharged and, applying the higher test, the outcome of the planning application would not have been any different if an appropriate assessment had been undertaken.

A new argument had surfaced on appeal to the Court of Appeal. It concerned a buffer zone. The claim was that Natural England had required a buffer zone of 20 metres around the proposed development, to protect the adjoining Windsor Forest and Windsor Great Park. However, permission had been given for a 15 metre zone, which, it was argued, meant that the judge at first instance, Lang J, had been wrong to conclude an assessment would have made no difference to the outcome.

Although Coulsen LJ had accepted Mr Hudson's argument that the planning permission was granted on the basis of a 20 metres buffer for the operational phase, he said that, even if the measurements were unclear, he was not persuaded that the accuracy of the measurements had anything to do with the absence of an 'appropriate assessment', or the possible outcome of the planning application, if there had been an 'appropriate assessment'.

3.3.7.3 Noise assessments

Where a proposed or new development presents a risk of noise pollution, to ensure planning consents and permissions it is usual to liaise with the LPA(s) and the Environment Agency. The National Planning Policy Framework (NPPF) (HCLG, 2021) sets out government's planning policies for England. All LPAs must have regard to the principles in the NPPF, particularly sustainable development, when producing a local plan.

Paragraph 174 of the NPPF states the requirements in terms of noise

> 174. Planning policies and decisions should contribute to and enhance the natural and local environment by
>
> ....
>
> e) preventing new and existing development from contributing to, being put at unacceptable risk from, or being adversely affected by, unacceptable levels of soil, air, water or noise pollution or land instability. Development should, wherever possible, help to improve local environmental conditions such as air and water quality, taking into account relevant information such as river basin management plans; and ...

Paragraph 185 provides additional detail

> 185. Planning policies and decisions should also ensure that new development is appropriate for its location taking into account the likely effects (including cumulative effects) of pollution on health, living conditions and the natural environment, as well as the potential sensitivity of the site or the wider area to impacts that could arise from the development. In doing so they should:
>
> (*a*) mitigate and reduce to a minimum, any potential adverse impacts resulting from noise from new development – and avoid noise giving rise to significant adverse impacts on health and the quality of life [See Explanatory Note to the Noise Policy Statement for England (Defra, 2010)]
> (*b*) identify and protect tranquil areas which have remained relatively undisturbed by noise and are prized for their recreational and amenity value for this reason and
> (*c*) limit the impact of light pollution from artificial light on local amenity, intrinsically dark landscapes and nature conservation.

Paragraph 188 refers to the overlap between planning requirements and other regulatory regimes, it states

> 188. The focus of planning policies and decisions should be on whether proposed development is an acceptable use of land, rather than the control of processes or emissions (where these are subject to separate pollution control regimes). Planning decisions should assume that these regimes will operate effectively. Equally, where a planning decision has been made on a particular development, the planning issues should not be revisited through the permitting regimes operated by pollution control authorities.

3.3.7.3.1 Noise management plan. The Environmental Permitting (England and Wales) Regulations 2016 (as amended) requires that all installations prevent or reduce to a minimum all pollutants which arise from them. This includes noise and vibrations. The size and location of a 'development', to use the word in a planning context, may cause adverse noise impacts to sensitive receptors during and/or after construction when it is in operation.

It is therefore advisable to produce a noise management plan to accompany a planning application for permission to operate an installation, as well as for the building stage. The plan should demonstrate the 'best available techniques'.

Proposed activities which may produce noise pollution should be identified prior to construction. Data should be provided in the plan in the form of predicted maximum and predicted average noise levels. These data can be compared directly with World Health Organization and current British standards in conjunction with available background noise data.

3.3.7.3.2 Measuring noise and vibration impacts. The UK government's planning policy guidance provides helpful clarification of how impacts can be characterised and translated into easily understood measures of impact. The standards required to be met depend on the nature of the sound and the acoustic environment within which the sound is perceived. The 2022 government guidance 'Noise and vibration management: environmental permits' (Environment Agency, 2022) should also be referred to, as it replaces the prior H3 guidance.

The planning policy guidance recommends that, when determining the impact of sounds, the LPA's plan-making and decision-taking should take account of the acoustic environment, and consider whether or not

- a significant adverse effect is occurring or likely to occur
- an adverse effect is occurring or likely to occur
- a good standard of amenity can be achieved.

This includes (in line with the Explanatory Note of the Noise Policy Statement for England (Defra, 2010)) identifying whether the overall effect of the noise exposure (including, wherever applicable, the impact during the construction phase) is, or would be, above or below the significant observed adverse effect level and the lowest observed adverse effect level for the given situation. This may require obtaining experienced specialist assistance.

The planning policy guidance also stipulates the appropriate actions in accordance with the likely response to noise (DLUHC, 2019b).

3.3.7.4 Flood risk assessments

The Environment Agency should be contacted if work is needed to be done near a sea defence, river or flood defence. A flood risk assessment may be needed before the planning authority will grant planning permission for the development to go ahead. These assessments should usually be completed by a professional flood risk assessor at the design stage of the proposed development, so that any changes can be incorporated in any architectural and engineering designs.

A flood risk assessment may be required as part of the planning application. An environmental permit may also be needed. For more information about flood risks, environmental permits and planning requirements, see Sections 2.6.4 and 2.12.6 in Chapter 2, and Sections 6.6 and 6.11 in Chapter 6.

3.4. The environmental statement (ES)

An ES should be completed by relevant experts and be accompanied by a statement from the developer providing evidence of their qualifications and experience. The applicant is responsible for the ES. It is attached to the planning application. It is the primary written submission.

As soon as the applicant for the proposed development notifies the relevant planning authority that they intend to submit an ES, the planning authority should notify any relevant consultation bodies, such as Natural England, the Office for Environmental Protection and the Marine Management Organisation. The planning authority will remind the relevant consultation bodies that they must make any relevant non-confidential information available to the applicant (Environmental Information Regulations 2004). The planning authority should also send the names and addresses of such consultation bodies to the applicant (regulation 17).

3.4.1 ES – scope opinion and screening opinion

Before an application is made to the LPA the applicant may ask for a formal scope opinion as to what information should be included in the ES (regulations 15 and 16 of the TCPEIA). This is called a 'scope opinion'. The LPA must ask for additional information if it does not have sufficient information to give an opinion (regulation 15(3)).

The LPA must also consult with relevant consultation bodies, such as Natural England, the Office for Environmental Protection and the Marine Management Organisation, before giving the scoping opinion. It has five weeks within which to provide the scoping opinion to the applicant. If the LPA fails to provide a scoping opinion, a request can be made to the Secretary of State (DLUHC, 2020).

The request for a scope opinion can be made at the same time as a request for a screening opinion. All these documents must be available to the public for a period of two years (regulation 28 of the TCPEIA).

If the developer does not agree with the LPA's scoping decision, they cannot appeal to the Secretary of State (regulation 15(17) of the TCPEIA 2017). Instead, there may be a judicial review of the decision on the basis that there has been a breach of the TCPEIA 2017 regulations.

3.4.2 Information required in the ES

Schedule 4 and regulation 18 of the TCPEIA and regulation 14 of the IPEIA set out the type of information that the applicant must give in the ES.

The ES has no prescribed form, but it must include

- a description of the whole project, including any demolition works
- a description of the main characteristics of the operational phase of the development (in particular any production process), for instance, energy demand and the energy used, the nature and quantity of the materials and natural resources (including water, land, soil and biodiversity) used
- an estimate, by type and quantity, of expected residues and emissions (e.g. water, air, soil and subsoil pollution, noise, vibration, light, heat and radiation), and quantities and types of waste produced during the construction and operation phases

- a description of the significant environmental effects of the project on particular things (e.g. flora, fauna, landscape, water, air and climate)
- data to identify and assess the principal effects of the project on the environment
- a statement of the proposed measures to be taken to avoid, reduce or alleviate these effects
- a summary in non-technical language.

The ES should set out the direct effects and any indirect, secondary, cumulative, trans-boundary, short-term, medium-term and long-term, permanent and temporary, positive and negative effects of the development.

The information provided may contain complex scientific data and specialised terminology. Therefore, it should be written in plain English so that it can be understood by non-experts, decision-makers and the general public.

3.4.3 The ES and the EIA

The statement should also be written bearing in mind that the EIA (i.e. the assessment made by the LPA or Planning Inspectorate (Secretary of State)) (part 1, section 4 of the TCPEIA)

> (2) … must identify, describe and assess in an appropriate manner, in light of each individual case, the direct and indirect significant effects of the proposed development on the following factors
>
> (*a*) population and human health
> (*b*) biodiversity, with particular attention to species and habitats protected under [*legislation*]
> (*c*) land, soil, water, air and climate
> (*d*) material assets, cultural heritage and the landscape
> (*e*) the interaction between the factors referred to in sub-paragraphs (a) to (d).

Practically, this would include taking into account

(*a*) the magnitude and spatial extent of the impact (for example geographical area and size of the population likely to be affected)
(*b*) the nature of the impact
(*c*) the transboundary nature of the impact
(*d*) the intensity and complexity of the impact
(*e*) the probability of the impact
(*f*) the expected onset, duration, frequency and reversibility of the impact
(*g*) the cumulation of the impact with the impact of other existing and/or approved development
(*h*) the possibility of effectively reducing the impact.

The effects to be considered must include the operational effects of the proposed development, wherever the proposed development will have operational effects (section 4 (3)).

The expected significant effects arising from the vulnerability of the proposed development to major accidents or disasters that are relevant to that development, as well as any other significant effects, should be identified, described, and assessed.

Box 3.4 *Finch On Behalf of the Weald Action Group, R (On the Application Of) v Surrey County Council & Ors* [2022] EWCA Civ 187

This was an appeal against a High Court decision. The issue was whether it was unlawful for Surrey County Council not to require the EIA for a commercial crude-oil extraction project to include an assessment of the impacts of greenhouse gas emissions resulting from the eventual use of the refined products of that oil as fuel. The parties agreed that it was common ground that an ES should assess both the direct and indirect effects of the development for which planning permission was sought that are likely to be significant; and that 'indirect effects' must be effects that the development itself has on the environment.

The Court of Appeal agreed with the High Court that Surrey County Council had not acted unlawfully. However, the Court of Appeal did not agree with the High Court that the assessment of greenhouse gas emissions from the future combustion of refined oil products at the development site was, as a matter of law, incapable of falling within the scope of the EIA for the planning application. The Court of Appeal held that the existence and nature of 'indirect' effects would always depend on the particular circumstances of the development under consideration, and that the need for a wider assessment of greenhouse gas emissions may sometimes be appropriate depending on the degree of connection between the development and its effects. However, the Court of Appeal decided that it was for Surrey County Council to establish whether the greenhouse gas emissions generated were properly to be regarded as 'indirect' effects of the proposed development. It was not the court's role in a claim for judicial review to substitute its own view for that of the planning authority (the Council) on this question.

3.4.4 Procedure for submission of an ES

The applicant should send the following to the planning authority.

- A planning application with two copies of the ES as well as all the documents that must normally accompany a planning application (regulation 19). The additional ES will be forwarded to the Secretary of State.
- The names of everyone to whom a copy of the ES has been sent or will be sent to, with the date on which the ES was sent, or served, where relevant.
- Enough copies of the ES for the planning authority to send to all consultation bodies that have not received a copy directly from the applicant and to all those likely to be affected.

The applicant should also make copies of the ES available to the public for free or at a reasonable cost (to reflect any printing and distribution costs).

3.4.5 Following submission of an ES to the planning authority

Once the ES together with the planning application has been received by the planning authority, it will check that all the information required by the regulations (e.g. regulation 4 of the TCPEIA and regulation 5 of the IPEIA) has been provided as well as any information required by schedule 4 of the regulations.

At this point, the planning authority is entitled to require the applicant to provide further information in writing, as well as evidence to support the ES (regulation 25). However, such requests should concern relevant matters about the main or significant environmental effects of the development (regulation 26 of the TCPEIA).

A copy of the ES and the planning application must be sent by the LPA to the Secretary of State within 14 days of its receipt.

A parish council in the LPA area may request, in writing, to be notified by the LPA of any relevant planning application, and any alteration to that application accepted by the authority if there is a relevant neighbourhood development plan for a neighbourhood area all or part of which falls within the authority's area (schedule 1 of the Town and Country Planning Act 1990 and articles 25 and 25A of the DMPO). The LPA must provide the information (section 61F of the Town and Country Planning Act 1990).

3.4.5.1 Consultation, publication and notification

Once the applicant's statement has been submitted, there must be consultation with all relevant bodies to assist the planning authority's assessment of the environmental impacts.

It may also be required to consult statutory consultees, such as, for example, other LPAs, the Canal & River Trust, the Gardens Trust, Historic England and Natural England (schedule 4 of the DMPO).

The Secretary of State can direct the LPA to carry out consultation in specific areas and on specific routes if the proposed development is near existing facilities (e.g. an airport, military installation or military explosive storage site).

The LPA can advise a developer of any directions on consultation in its area. These directions typically set out detailed maps of the areas involved. For example, the Town and Country Planning (Safeguarding Meteorological Sites) (England) Direction 2014 sets out who should be consulted when a proposed development may affect a meteorological technical site, and the Town and Country Planning (Safeguarded Aerodromes, Technical Sites and Military Explosives Storage Areas) Direction 2002 provides advice and sets out who should be consulted when a proposed development may affect an aerodrome, technical site or military explosives storage area.

3.4.5.2 Planning register

The ES and any related documents including screening or scoping opinions or direction must be placed on part I of the planning register, as soon as possible after publication.

3.4.5.3 Publicity requirements

The LPA must publicise the planning application and ES as set out in articles 15 and 16 of the DMPO. However, where the ES has been submitted to the LPA after submitting the planning application, it is the applicant who is responsible for publicising the application and ES. In addition, in such a case the applicant must ensure that the submission of the ES is accompanied by certificates stating that the publicity arrangements have been met (regulation 20 of the TCPEIA).

Where the land area of the development lies within the boundaries of two LPAs the required publicity activities must be carried out separately in each LPA area.

The LPA or the applicant, when responsible for publication, must take reasonable steps to serve a notice with the required information to those persons or bodies who otherwise would be unaware of the ES, but who may have an interest in, or be affected by, the proposed development.

3.4.5.4 Length of publication and responses

The planning application and the ES must be publicised by notices in the local press for 14 days, and onsite or near the site for 30 days or, where the development concerns a public right of way or the application is only for a technical details consent, for 21 days.

A parish council or neighbourhood forum must make representations within 21 days of notification of a planning application.

Where the planning application is for a major development not requiring an EIA application and ES, it must be publicised by notices placed in the local press and displayed onsite or near the site for 21 days, and served on owners/occupiers of adjoining land.

The LPA is required by various legislation, including the DMPO, to allow an extra day for publicity of applications for each public holiday.

Statutory consultees must respond to applications (section 54 of the Planning and Compulsory Purchase Act 2004). Responses from various consultation bodies and statutory consultees and the public needs to be received within the above stated periods of time.

Where the application is for a public service infrastructure development made on or after 1 August 2021, the period for the response by all concerned is only 18 days. However, a different period may be agreed in writing between the consultee and consultor.

Whichever the case, the relevant time periods for responses/comments must be set out in the publicity (i.e. the notices sent out, or placed in the local press or onsite).

3.4.5.5 Information on the LPA website

Where an EIA application is accompanied by an ES, an application is for a major development not requiring an EIA and ES, or an application is only for technical details consent, regulation 15(7) of the DMPO states that

> (7) The following information must be published on a website maintained by the local planning authority
>
> (*a*) the address or location of the proposed development

(*b*) a description of the proposed development
(*c*) the date by which any representations about the application must be made, which must not be before the last day of the period of 14 days beginning with the date on which the information is published
[*In the case of an EIA application accompanied by an ES, the latter must also include the date by which any representations about the application must be made. This must not be before the last day of the period of 30 days, beginning with the date on which the information is published.*]
(*d*) where and when the application may be inspected
(*e*) how representations may be made about the application and
(*f*) that, in the case of a householder or minor commercial application, if there is an appeal that proceeds by an expedited procedure, any representations made about the application will be passed to the Secretary of State, and there will be no opportunity to make further representations.

Where the applicant is responsible for publication of notices, the LPA must also place all submitted documents on its website, including any scope and screening documents.

3.4.5.6 Format and content of planning notices

A planning notice is displayed on or near the development site or published in a newspaper and served on the owners/occupiers of adjoining land or infrastructure managers by the LPA. The appropriate format and content for planning application notices apart from EIA applications is provided as a proforma in schedule 3 of the DMPO, and it can be copied.

3.5. Consideration and determination of EIAs and appeals

The LPA must consider the ES, any responses from consultation bodies and the public, and any relevant information before making a decision. The GOV.UK website application flowchart sets out each stage of the determination procedure and the timeframes (HMG, 2021b).

3.5.1 Factors required to be considered in applications or appeals

When considering whether planning permission or subsequent consent should be granted, whether determining an application or an appeal in relation to which an ES has been submitted, the relevant planning authority (the Secretary of State or a planning inspector) must

- examine the environmental information
- reach a reasoned conclusion on the significant effects of the proposed development on the environment, taking into account the examination and, where appropriate, the planning authority's own supplementary examination
- integrate that conclusion into the decision as to whether planning permission or subsequent consent is to be granted, and
- if planning permission or subsequent consent is to be granted, consider whether it is appropriate to impose monitoring measures.

Planning permission or subsequent consent for EIA development must not be granted unless the relevant planning authority (the Secretary of State or the inspector) is satisfied that the reasoned conclusion is up to date. It is up to date if, in the opinion of the relevant planning authority, it addresses the significant effects of the proposed development on the environment that are likely to arise as a result of the proposed development.

3.5.2 Notification of decision

The day after the submission of the ES and planning application is the start of the period of 16 weeks for the LPA to reach a determination on the application. This period continues even though the authority may need to correspond with the applicant requiring further information during this time. This period may be extended if both the LPA and the applicant agree to do so in writing (regulation 68(2) of the TCPEIA).

In cases where there is no statutory timescale in place, the decision of the relevant authority or the Secretary of State, depending on who is required to decide, must be taken within a reasonable period of time, from the date on which either has been provided with the environmental information (regulation 26 of the TCPEIA).

Once a decision has been made by the LPA, the Secretary of State or a planning inspector, they must inform the developer of the decision and provide the following information (regulation 29 of the TCPEIA).

- The right to challenge the decision and the procedures for doing this.
- If the decision is to grant planning permission or subsequent consent
  - the reasoned conclusion of the relevant planning authority or the Secretary of State on the significant effects of the development on the environment
  - any conditions to which the decision is subject which relate to the likely significant environmental effects of the development
  - a description of any features of the development and any measures to avoid, prevent, reduce and, if possible, offset likely significant adverse effects on the environment
  - any monitoring measures considered appropriate by the relevant planning authority or the Secretary of State.
- If the decision is to refuse planning permission or subsequent consent, the main reasons for the refusal.

3.5.2.1 LPA and Secretary of State to notify the decision to the public and relevant others

Under regulation 30(1) of the TCPEIA

(1) Where an EIA application is determined by an LPA, the authority must promptly

(*a*) inform the Secretary of State of the decision in writing
(*b*) inform the consultation bodies of the decision in writing
(*c*) inform the public of the decision, by local advertisement, or by such other means as are reasonable in the circumstances, and

(*d*) make available for public inspection at the place where the appropriate register (or relevant part of that register) is kept a statement containing
(i) details of the matter referred to in regulation 29(2).

Under regulations 30(2) and 30(3) of the TCPEIA

(2) Where an EIA application or appeal is determined by the Secretary of State or an inspector, the Secretary of State must

(*a*) notify the relevant planning authority of the decision, and
(*b*) provide the authority with such a statement as is mentioned in paragraph (1)(d).

(3) The relevant authority must, as soon as reasonably practicable after receipt of a notification under paragraph (2)(a), comply with sub-paragraphs (b) to (d) of paragraph (1) in relation to the decision so notified as if it were a decision of the authority.

3.5.2.2 Consent and monitoring

The relevant planning authority, the Secretary of State or inspector, as appropriate, must consider whether to impose monitoring measures. In doing so, they must (regulation 26 of the TCPEIA)

(*a*) if monitoring is considered to be appropriate, consider whether to make provision for potential remedial action
(*b*) take steps to ensure that the type of parameters to be monitored and the duration of the monitoring are proportionate to the nature, location and size of the proposed development and to the significance of its effects on the environment, and
(*c*) consider, to avoid duplication of monitoring, whether any existing monitoring arrangements carried out in accordance with the law in any part of the United Kingdom are more appropriate than imposing a monitoring measure.

3.6. Mitigation measures

Mitigation measures are supposed to be designed to reduce or remove any significant damaging effects of the development. If the LPA decides to impose such measure, they must clearly identify and explain what they are. It is insufficient for the LPA to state as a condition of granting planning permission that the submitted ES must be followed, unless the ES had stated what mitigations would be applied and had stated them fully, clearly and precisely, and they had fully covered all measures required by the LPA.

It is prudent for an LPA to state clearly and precisely what mitigation measures must be taken to avoid any disputes over the clarity of the measures. It is therefore advisable for the LPA to refer to sections of the ES, if it refers to the development's self-imposed mitigation measures, and, where necessary, elaborate, what more needs to be done or changed.

As well as, or in addition to, providing conditions to the granting of planning permission or consents, the LPA may decide to require that the developer enters into a legally binding agreement to carry out planning obligations.

3.6.1 Environment Act 2021 – biodiversity net gain (BNG)

In addition to current mitigation measures, section 98 of part 6 of the Environment Act 2021 refers to schedule 14 within that Act, which makes biodiversity gain a condition of obtaining planning permission. Similarly, section 99 refers to schedule 15, which makes biodiversity gain a condition of development consent for nationally significant infrastructure projects.

Schedule 14 of the Act required that a new section (section 90A) and a schedule 7A be inserted the Town and Country Planning Act 1990 so that an applicant seeking permission for a development will have to submit a biodiversity-gain plan demonstrating to the satisfaction of the LPA that, as per schedule 14 of the Environment Act 2021, the development will achieve a 10% biodiversity gain. That is, that the biodiversity value attributable to the development will exceed the pre-development biodiversity value of the onsite habitat by at least the relevant percentage, which is currently 10%. However, Simmonds *et al.* (2022) found that, even where compensatory gains are absolute, the arbitrary determination of how much gain is required per unit of loss (e.g. England's policy requires a minimum net gain of 10%), may mean that the gains necessary to help achieve desired conservation outcomes such as a 2030 or 2050 target are not achieved. However, the Secretary of State may, by regulation, change the relevant percentage to a higher or lower value.

The land that the BNG refers to concerns land that is to be developed or maintained for 30 years (section 90A of the Town and Country Planning Act 1990 (TCPA)). Under section 100 of the Environment Act 2021 a register of these 'net gain sites' will be created, and the local authority or designated body must ensure that the gain itself is in place for 30 years. If it is not possible to achieve the net gain on the site described in the application itself, the applicant may propose to improve the biodiversity of an alternative area of land, with any enhancements recorded in the risk register. If that is not possible, then under section 101 of the Environment Act 2021 the applicant may purchase what will be known as 'biodiversity credits' for the purpose of meeting the BNG objective referred to in schedule 7A of the Town and Country Planning Act 1990, or schedule 2A of the Planning Act 2008 for major infrastructure projects. This is provided the relevant system has been put in place.

However, Zu Ermgassen *et al.* (2021) found that, although there are ambitious commitments to monitoring and implementing offsetting measures, little attention has been paid to ensuring the delivery of habitats within developer-owned land. Research by Simpson *et al.* (2022) shows that even a target of no net loss is insufficient to mitigate the impacts on a range of closely related habitats and species.

3.6.1.1 BNG mitigation process

Developers' proposals will have to follow a four-step mitigation so that the BNG of a site can be determined by the relevant planning authority. This process, which is already in use by planning authorities to assess environmental impacts, is supposed to ensure development plans can move through the planning process swiftly and meet all the legal requirements. However, the Environment Act 2021 requires more certainty in terms of what is required.

The first step in a mitigation process is avoidance, which requires developers to avoid taking action that may cause harm. This means altogether avoiding developing locations that are particularly rich in biodiversity. However, planning permission applicants may purchase biodiversity credits.

The second step is minimisation, which involves preliminary proactive measures to reduce the time, extent, impact and intensity of the development.

The third stage is onsite restoration, which requires the developer to enhance or restore habitats after the development is completed. This step is crucial if avoidance and minimisation were not possible in the first instance.

The final step, offsetting, concerns any measures needed to compensate for the adverse impacts of the development, after the previous mitigation stages have been addressed. However, strict enforcement of compliance with sanctions for non-compliers is needed to address low offsetting compliance rates (Gray and Shimshack, 2011).

3.6.1.2 Delivery and implementation of BNG

There are questions regarding when the BNG should be delivered. The detailed outcome of the government consultation, which 22 January 2022 consultation paper set out proposals on the detail of implementation of mandatory BNG, and which consultation closed on 5 April 2022, indicates that biodiversity gains can be delivered alongside other enhancement or mitigation measures (Defra, 2023). However, they will need to be separately accounted for and recorded in the register (Gateley, 2022).

The revised British Standard on biodiversity net gain and development (BS 8683:2021) (BSI, 2021a) does not cover the actual delivery of BNG, but it provides a consistent and structured process for designing and implementing BNG based on UK-wide good practice (BSI, 2021b). However, it is not law, and so it does not have to be followed.

3.6.1.3 BNG costs

Although the BNG regulatory provisions are not yet scheduled to be brought into force, to meet the requirements of the Environment Act 2021, when they are, developers will have to factor in and accommodate the new minimum 10% BNG requirement and engage with the LPA from the conception of the project. The latter will need to demonstrate that the BNG process has been thoroughly complied with if a dispute arises. It has been argued that the use of mediation is particularly beneficial for resolving planning and environment-related issues (Agapiou, 2018).

Planning authorities may require developers to provide a minimum 10% BNG now, given provisions 174, 179 and 180 of the National Planning Policy Framework (HCLG, 2021). The planning authorities would then review biodiversity submissions and monitor implementation. This process would therefore involve extra work for all the parties concerned. Furthermore, planning authorities will be able to charge developers for using council-owned land as off-site gain locations (Local Government Association, 2023). Therefore, when the regulatory provisions are brought into effect, which is unlikely to be before 2023, they are bound to impact on an assessment of the cost viability of a construction project.

3.6.2 Environment Act 2021 – nature recovery strategies

Sections 104 to 107 of the Environment Act 2021 require councils in England to create a local nature recovery strategy for areas to be decided by the Secretary of State. These space-specific strategies will be prepared and published by the relevant authority responsible for the area – a local authority, Mayor of London, Natural England, the Broads Authority or a National Park authority.

Section 106 clearly and simply states that the content of the local nature recovery strategy includes a statement of biodiversity priorities for the strategy area, and either a local habitat map for the whole strategy area or two or more local habitat maps which together cover the whole strategy area. The statement of biodiversity priorities referred to in subsection (1)(a) must include: a description of the opportunities for recovering or enhancing biodiversity, in terms of habitats and species, in the strategy area; the priorities, in terms of habitats and species, for recovering or enhancing biodiversity; and proposals as to potential measures relating to those priorities.

The strategies will highlight and prioritise opportunities for improving habitats in the most valuable wildlife areas in the country. The resulting strategy maps will offer opportunities for developers to improve the local environment. However, Smith *et al.* (2021) argue there is a need for 'national-level guidance and data for species, to support existing networks of local environmental record centres and to create a national framework to inform local action'.

3.6.3 Environment Act 2021 – conservation covenants

As the 10% BNG must be maintained for 30 years, part 7 of the Environment Act 2021 introduces conservation covenants to assist such compliance. These covenants are a type of contract which binds and runs with the land. Potentially, they can indefinitely restrict an owner's use of their land and access to ecosystems (Archibald *et al.*, 2021).

Developers will be required by the planning authority, as a condition of planning permission, to enter into a conservation covenant to ensure that any future owners of the development or the land itself retain biodiverse areas. Section 117(3) of the Environment Act 2021 defines the covenant as having a conservation purpose if its purpose is

(*a*) to conserve the natural environment of land or the natural resources of land
(*b*) to conserve land as a place of archaeological, architectural, artistic, cultural or historic interest, or
(*c*) to conserve the setting of land with a natural environment or natural resources or which is a place of archaeological, architectural, artistic, cultural or historic interest.

Therefore, in practice this could include a range of obligations, such as maintaining habitats on site or preventing certain actions. Failure to comply with the Environment Act 2021 could result in an injunction or a breach of contract (Lindsay, 2016).

3.6.4 The transition

Many of the requirements of the Environment Act 2021 are still to be brought into effect by regulation. Until then, the requirements do not constitute law that must be obeyed. The

Environment Act 2021 (Commencement No. 2 and Saving Provision) Regulations 2022 set out the dates for the commencement of sections of the Environment Act 2021, but it did not provide dates for all of them.

Sections 1 to 15 provisions of the Environment Act 2021 concern targets and environmental improvement plans. Part 6 sections 104 to 108 (local nature recovery strategies) came into force on 24 January 2022, and part 7 (conservation covenants) came into force on 30 September 2022.

At the time of writing, no date has been set for the sections relating to BNG (sections 98 to 103) to come into force. However, as explained above, the National Planning Policy Framework already requires an element of biodiversity net gain. Therefore, developers cannot ignore what is the start of a new age of addressing nature conservation issues, both urban and rural.

3.7. Challenging the LPA's decision – judicial review

The planning application applicant or the developer, if they are not the applicant, or an interested party who shows they have sufficient interest in the matter may challenge the LPA's EIA by way of judicial review, if the argument is that the law has been broken.

However, a judicial review is a last resort, and is an expensive process. Where possible, other legitimate means, such as negotiation, should be exhausted first.

A challenge may be made to the screening decision made by the Secretary of State.

Box 3.5 Swire, *R (On the Application Of) v Secretary of State for Housing, Communities and Local Government* [2020] EWHC 1298 (Admin)

This was a judicial review challenge by a local resident to a screening direction made by the Secretary of State. The matter concerned a development proposal to demolish existing structures and build up to 20 residential units in the Kent Downs area of Outstanding National Beauty. The Secretary of State's screening direction was similar to the original one issued by the LPAs. However, the Secretary of State's direction said that the proposed development was an EIA development because it was unlikely to have a significant effect on the environment, as the LPA had proposed conditions in the outline application decision that were sufficient to address any risks.

However, the High Court decided there was insufficient evidence to support this conclusion. The decision that an EIA was not required was therefore quashed (cancelled).

References

Statutes

Environment Act 2021
Planning Act 2008
Planning and Compulsory Purchase Act 2004
The Town and Country Planning Act 1990
Wildlife and Countryside Act 1981

Regulations

Environmental Permitting (England and Wales) Regulations 2016 (as amended)
The Conservation of Habitats and Species (Amendment) (EU Exit) Regulations 2019
The Conservation of Habitats and Species Regulations 2010
The Conservation of Habitats and Species Regulations 2017 (as amended)
The Electricity Works (Environmental Impact Assessment) (England and Wales) Regulations 2000
The Environment Act 2021 (Commencement No. 2 and Saving Provision) Regulations 2022
The Environmental Information Regulations 2004
The Infrastructure Planning (Environmental Impact Assessment) Regulations 2017
The Marine Works (Environmental Impact Assessment) Regulations 2007
The Offshore Oil and Gas Exploration, Production, Unloading and Storage (Environmental Impact Assessment) Regulations 2020
The Water Resources (Environmental Impact Assessment) (England and Wales) Regulations 2003
Town and Country Planning (Environmental Impact Assessment) (Wales) Regulations 2017
Town and Country Planning (Environmental Impact Assessment) Regulations 2017

Directives

The Habitats Directive 92/43/EEC
The Birds Directive 2009/147/EC

Directions

Town and Country Planning (Safeguarded Aerodromes, Technical Sites and Military Explosives Storage Areas) Direction 2002
Town and Country Planning (Safeguarding Meteorological Sites) (England) Direction 2014

Orders

The Town and Country Planning (Development Management Procedure) (England) Order 2015
The Town and Country Planning (General Development Procedure) (Amendment) (England) Order 2010

Case law

Berkley v Secretary of State for the Environment and others [2000] UKHL 36
Finch On Behalf of the Weald Action Group, R (On the Application Of) v Surrey County Council & Ors [2022] EWCA Civ 187
Hudson, R (On the Application Of) v Royal Borough of Windsor and Maidenhead & Ors [2021] EWCA Civ 592
R (On the Application of Champion) v North Norfolk District Council & Anor [2015] UKSC 52
Swire, R (On the Application Of) v Secretary of State for Housing, Communities and Local Government [2020] EWHC 1298 (Admin)

Journals

Agapiou A (2018) The potential of mediating planning and environmental disputes in England and Wales. *Proceedings of the Institution of Civil Engineers – Management, Procurement and Law* **171(3)**: 91–92. 10.1680/jmapl.18.00006.
Archibald CL, Dade MC, Sonter LJ *et al.* (2021) Do conservation covenants consider the delivery of ecosystem services? *Environmental Science & Policy* **115**: 99–107. 10.1016/j.envsci.2020.08.016.

Gray W and Shimshack J (2011) The effectiveness of environmental monitoring and enforcement: a review of the empirical evidence. *Review of Environmental Economics and Policy* **5**: 3–24. 10.1093/reep/req017.

Lindsay B (2016) Legal instruments in private land conservation: the nature and role of conservation contracts and conservation covenants. *Restoration Ecology* **24(5)**: 698–703. 10.1111/rec.1239.

Simmonds JS, von Hase A, Quétier F *et al.* (2022) Aligning ecological compensation policies with the Post-2020 Global Biodiversity Framework to achieve real net gain in biodiversity. *Conservation Science and Practice* **4**: e12634. 10.1111/csp2.12634.

Simpson KH, de Vries FP, Dallimer M, Armsworth PR and Hanley N (2022) Ecological and economic implications of alternative metrics in biodiversity offset markets. *Conservation Biology* **36(5)**: e13906. 10.1111/cobi.13906.

Smith RJ, Cartwright SJ, Fairbairn SC *et al.* (2021) Developing a nature recovery network using systematic conservation planning. *Conservation Science and Practice* **4**: e578. 10.1111/csp2.578.

Zu Ermgassen SOSE, Marsh S, Ryland K *et al.* (2021) Exploring the ecological outcomes of mandatory biodiversity net gain using evidence from early-adopter jurisdictions in England. *Conservation Letters* **14(6)**: e12820. 10.1111/conl.12820.

Websites

BSI (British Standards Institution) (2021a) BS 8683:2021. Process for designing and implementing Biodiversity Net Gain. Specification. BSI, London.

BSI (2021b) New British Standard sets out requirements for the implementation of Biodiversity Net Gain in development projects. https://www.bsigroup.com/en-GB/about-bsi/media-centre/press-releases/2021/august/new-british-standard-sets-out-requirements-for-the-implementation-of-biodiversity-net-gain-in-development-projects (accessed 09/03/2023).

CIEEM (Chartered Institute of Ecology and Environmental Management) (2021) Brexit changes to the Habitats Regulations for England and Wales. https://cieem.net/brexit-changes-to-the-habitats-regulations (accessed 09/03/2023).

Defra (Department for Environment, Food and Rural Affairs) (2010) Noise Policy Statement for England. https://www.gov.uk/government/publications/noise-policy-statement-for-england (accessed 09/03/2023).

Defra (2023) Consultation on Biodiversity Net Gain regulations and implementation. https://www.gov.uk/government/consultations/consultation-on-biodiversity-net-gain-regulations-and-implementation (accessed 09/03/2023).

Defra and Natural Resources Wales (2021) Habitats regulations assessments: protecting a European site. https://www.gov.uk/guidance/habitats-regulations-assessments-protecting-a-european-site (accessed 09/03/2023).

DLUHC (Department for Levelling Up, Housing and Communities) (2019a) Before submitting and application. https://www.gov.uk/guidance/before-submitting-an-application (accessed 09/03/2023).

DLUHC (2019b) Noise. https://www.gov.uk/guidance/noise–2#full-publication-update-history (accessed 09/03/2023).

DLUHC (2020) Environmental Impact Assessment. https://www.gov.uk/guidance/environmental-impact-assessment#statutory-consultation-bodies (accessed 09/03/2023).

Environment Agency (2022) Noise and vibration management: environmental permits. https://www.gov.uk/government/publications/noise-and-vibration-management-environmental-permits/noise-and-vibration-management-environmental-permits (accessed 09/03/2023).

European Union (1992) Council Directive 92/43/EEC of 21 May 1992 on the conservation of natural habitats and of wild fauna and flora. https://www.legislation.gov.uk/eudr/1992/43/contents# (accessed 09/03/2023).

European Union (2009) Directive 2009/147/EC of the European Parliament and of the Council of 30 November 2009 on the conservation of wild birds (codified version). https://www.legislation.gov.uk/eudr/2009/147# (accessed 09/03/2023).

Gateley (2022) Biodiversity net gain regulations consultation. Gately Legal, Guildford. https://gateleyplc.com/insight/quick-reads/biodiversity-net-gain-regulations-consultation (accessed 09/03/2023).

HCLG (Ministry of Housing, Communities and Local Government) (2021) National Planning Policy Framework. https://www.gov.uk/government/publications/national-planning-policy-framework–2 (accessed 09/03/2023).

HMG (His Majesty's Government) (2017) Explanatory Memorandum to the Conservation of Habitats and Species Regulations 2017. https://www.legislation.gov.uk/uksi/2017/1012/memorandum/contents (accessed 09/03/2023).

HMG (2021a) Schedule 2 criteria and thresholds. https://assets.publishing.service.gov.uk/government/uploads/system/uploads/attachment_data/file/630689/eia-thresholds-table.pdf (accessed 09/03/2023).

HMG (2021b) EIA application flowchart. https://assets.publishing.service.gov.uk/government/uploads/system/uploads/attachment_data/file/630687/eia-flow2.pdf (accessed 09/03/2023).

JNCC (Joint Nature Conservation Committee) (2019) Ramsar Convention. https://jncc.gov.uk/our-work/ramsar-convention (accessed 09/03/2023).

JNCC (2021a) SACs in England. https://sac.jncc.gov.uk/site/england (accessed 09/03/2023).

JNCC (2021b) Special protection areas. https://jncc.gov.uk/our-work/special-protection-areas-overview (accessed 09/03/2023).

Local Government Association (2023) Biodiversity Net Gain FAQs. https://www.local.gov.uk/pas/topics/environment/biodiversity-net-gain-local-authorities/biodiversity-net-gain-faqs#bng-alongside-other-mitigation-and-benefits-additionality-stacking-and-bundling-natural-capital (accessed 09/03/2023).

Natural England (2019) Pre-submission screening service: advice on planning proposals affecting protected species. https://www.gov.uk/guidance/pre-submission-screening-service-advice-on-planning-proposals-affecting-protected-species (accessed 09/03/2023).

Natural England (2022) European protected species: apply for a mitigation licence (A12). https://www.gov.uk/government/publications/european-protected-species-apply-for-a-mitigation-licence (accessed 09/03/2023).

Natural England (2022) Protected species: apply for a mitigation licence (A05 and A05a). https://www.gov.uk/government/publications/protected-species-apply-for-a-mitigation-licence-a05-and-a05a (accessed 09/03/2023).

Natural England and Defra (Department for Environment, Food and Rural Affairs) (2022) Wildlife licences: when you need to apply. https://www.gov.uk/guidance/wildlife-licences (accessed 09/03/2023).

Natural England (2023) MAGIC. Interactive mapping at your fingertips. https://magic.defra.gov.uk (accessed 09/03/2023).

Baker F and Charlson J
ISBN 978-0-7277-6645-8
https://doi.org/10.1680/elsc.66458.107

Chapter 4
Contaminated and brownfield land

Jennifer Charlson

4.1. Introduction

This chapter details the legal frameworks regarding contaminated land and brownfield land. Both frameworks aim for the redevelopment of land but use different approaches.

Remediation of contaminated land can be mandated by part IIA of the Environmental Protection Act 1990. The Town and Country Planning (Brownfield Land Register) Regulations 2017 require local authorities to publicise brownfield sites with the added incentive of planning permission in principle. However, there is no obligation for the brownfield sites to be redeveloped.

Remediation of contaminated land can be mandatory, whereas redevelopment of brownfield land is voluntary. Therefore, it is important to distinguish between them.

4.1.1 Contaminated land

The legal regime for contaminated land was introduced by part IIA of the Environmental Protection Act 1990. The definition of contaminated land is detailed. The inspection and enforcement system is outlined, and appropriate (class A and class B) responsible persons are identified.

A risk-based approach is adopted for the assessment of contaminated sites. The Environment Agency's land contamination risk management guidance is signposted.

Cases concerning remediation notices served on developers and liability for contaminated land following statutory transfer schemes are described, and the contaminated land regime is critiqued.

4.1.2 Environmental Damage (Prevention and Remediation) (England) Regulations 2015

The Environmental Damage (Prevention and Remediation) (England) Regulations 2015 are explained. The operator's mandatory requirements are outlined and the regulator's enforcement options detailed.

4.1.3 Brownfield land

Brownfield land is defined in the National Planning Policy Framework (HCLG, 2021).

The Town and Country Planning (Brownfield Land Register) Regulations 2017 mandated local planning authorities (LPAs) to prepare, maintain and publish a brownfield land register. An LPA is required to publish details about sites, consult before sites can be granted permission in principle, and regularly review the suitability of all relevant sites.

There is no right of appeal when an LPA decides not to enter a site in part 2 of a brownfield land register. Following a grant of permission in principle, the site must receive a grant of technical details consent for a grant of planning permission.

HM Treasury provides funding to bring brownfield land into use. CPRE (formerly the Campaign to Protect Rural England) reports annually on the status and potential of brownfield land in England.

4.2. Contaminated land

England has a proud industrial history, but one of the legacies is contaminated land.

4.2.1 Environmental Protection Act 1990 – part IIA

The purpose of part IIA of the Environmental Protection Act 1990 is to identify and remediate contaminated land in England that poses an unacceptable risk to human health or the environment and which has not been remediated voluntarily.

4.2.2 Contaminated land – definition

Contaminated land is defined in section 78A(2) of the Environmental Protection Act 1990 as

> (2) … any land which appears to the local authority in whose area it is situated to be in such a condition, by reason of substances in, on or under the land, that
>
> (*a*) Significant harm is being caused or there is a significant possibility of such harm being caused, or
> (*b*) Pollution of controlled waters is being, or is likely to be, caused.

Part IIA takes a risk-based approach to defining contaminated land. For a relevant risk to exist there needs to be one or more contaminant–pathway–receptor linkages (‘contaminant linkage’) by which a relevant receptor might be affected by the contaminants.

4.2.3 Inspection and enforcement

The Secretary of State for the Environment, Food and Rural Affairs issued statutory guidance (Defra, 2012), in accordance with section 78W, which is binding on enforcing authorities. Local authorities are required to devise and implement an inspection strategy, but no deadline was imposed. Therefore, local authorities have made slow progress in inspecting their areas, resulting in a lack of data on the extent of contaminated land in England (Fogleman, 2014a).

Once land has been determined as contaminated land, the enforcing authority must consider how it should be remediated and, where appropriate, must issue a remediation notice to require remediation. The enforcing authority may be the local authority which determined the land, or the Environment Agency, which takes on responsibility for ‘special sites’. The rules regarding special sites and issuing of remediation notices are set out in the Contaminated Land (England) Regulations 2006.

Enforcing authorities have the power to carry out remediation (section 78N). Guidance on the extent to which the enforcing authority should seek to recover the costs of remediation which it has carried out is issued under section 78P (Defra, 2012).

4.2.4 Appropriate responsible person

The appropriate person to bear responsibility for remediation is determined by section 78F. The person (class A person) primarily liable for remediation costs is the one who caused or knowingly permitted the contamination. If no class A person can be found, the authority usually seeks to identify the owners or occupiers of land (class B person).

The 'sold with information' test may exclude from liability sellers who, although they have caused or knowingly permitted the presence of a significant contaminant in, on or under some land, have sold that land in circumstances where it is reasonable that the purchaser should bear the liability for remediation of the land. Similarly, the 'introduction of pathways or receptors' test may exclude from liability those who would otherwise be liable solely because of the subsequent introduction by others of the relevant pathways or receptors in the significant contaminant linkage.

A person on whom a remediation notice is served may appeal against the notice (section 78L). However, a person who, without reasonable excuse, fails to comply with a remediation notice served by an enforcing authority is guilty of an offence (section 78M).

4.2.5 Land contamination risk management

A risk-based approach is adopted to the assessment of contaminated sites. The overarching principle requires a complete 'contaminant linkage' to be present, consisting of all the following

- source – the contaminant(s) presenting a risk to human health or the environment
- pathway – the route by which a contaminant comes into contact with a receptor
- receptor – the entity that could be harmed through contact with a contaminant.

The Environment Agency has published guidance, which it expects to be followed, on how to assess and manage the risks from land contamination (Environment Agency, 2021). The land contamination risk management guidance comprises four guides

- before you start
- stage 1 risk assessment
- stage 2 options appraisal
- stage 3 remediation and verification.

For land contamination, a competent person with a recognised relevant qualification, sufficient experience in dealing with the type(s) of pollution and membership of a relevant professional organisation must be employed.

4.2.6 Remediation notices served on developers

Developers have been served remediation notices, as illustrated by the cases concerning Circular Facilities (London) Ltd and Crest Nicholson Residential Ltd detailed below.

4.2.6.1 Circular Facilities (London) Ltd

Box 4.1 *Circular Facilities (London) Ltd v Sevenoaks District Council* [2005] EWGC 865 (Admin)

Circular Facilities (London) Ltd (CFL) developed a residential estate in Tonbridge, Kent, in the early 1980s. The site was discovered by Sevenoaks District Council (SDC) to be causing significant risk of harm to the residents as a result of high levels of methane and carbon dioxide seeping from pits filled with now decomposing organic material. The investigation revealed that the 'pathway' by which the 'contaminant' reached the 'receptor' was via service points and cracks or fissures in the concrete floors of the houses.

In 2002, SDC served a remediation notice on CFL as a class A person. CFL appealed against the remediation notice. This was the first appeal to reach the High Court arising out of part IIA of the Environmental Protection Act 1990, in connection with a contaminated land remediation notice served by a local authority. It was agreed that the land was contaminated, but the issue was whether CFL was an 'appropriate person'. The case was confidentially settled out of court and the notice revoked.

4.2.6.2 St Leonard's Court

St Leonard's Court is a residential estate near St Albans, Hertfordshire. The site had previously been the location of a chemical works, from the 1950s until about 1980, which was subsequently purchased by a developer. Bromate and bromide were discovered to have leached into the soil, contaminating the water source. In 2002, the site was identified as contaminated land, and the matter was referred to the Environment Agency to take enforcement action.

The Environment Agency served remediation notices on both the former chemical works operator (Redland Minerals Ltd) and the developer (Crest Nicholson Residential Ltd).

Box 4.2 *Redland Minerals Ltd, R (On the Application Of) v Secretary of State for Environment, Food and Rural Affairs* [2010] EWHC 913 (Admin)

In an appeal against one of the remediation notices, the judge concluded

> [....] not only that a manufacturer of chemicals had caused chemical contaminants to enter soil and groundwater, but that a developer that subsequently removed buildings and hardstandings from the site during its redevelopment for housing, had caused the chemicals to enter the groundwater due to rain falling directly on the ground and accelerating the entry of contaminants to lower levels in the ground.

Box 4.3 *Crest Nicholson Residential Ltd, R (On the Application Of) v Secretary of State for Environment, Food and Rural Affairs & Ors* [2010] EWHC 1561 (Admin)

The developer (Crest Nicholson Residential Ltd) had carried out some testing on the site and had identified bromide in the upper soil, but did not appreciate how deeply that bromide had penetrated the soil and into the aquifer.

The judge found both parties, the chemical works (Redland Minerals Ltd) and developer (Crest Nicholson Residential Ltd), liable and apportioned liability between them.

4.2.7 Liability for contaminated land following statutory transfer schemes

The liability of National Grid Gas plc and Powys County Council for contaminated land following statutory transfer schemes is described in the following cases.

4.2.7.1 National Grid Gas plc

Box 4.4 *R (On the Application of National Grid Gas plc (formerly Transco plc)) v Environment Agency* [2007] UKHL 30

In 2001, residents of land near Bawtry, Doncaster, discovered a pit with coal tar, which was a by-product of coal gas-making processes, in their back gardens. The site was subsequently determined to be contaminated land. As coal tar is potentially harmful to health, the Environment Agency remediated the site and then sought to recover the cost.

In this case, the House of Lords identified three possible payers: the polluters, the present owners of the residences beneath whose gardens the coal tar was buried and public funds. The House of Lords then considered whether liability for remediation could be transferred notwithstanding the fact that liability was created after statutes that transferred liability from statutory predecessors. The site had been a gas works operated by statutory predecessors to National Grid Gas plc (NGG) from approximately 1915, and was later developed for housing in the 1960s.

It was held that NGG had not itself caused or knowingly permitted the presence of the contamination, as it had only come into existence 20 years after the site had been sold for housing. It was deemed that past liability, for the acts of the predecessor, had not been created by the Environmental Protection Act 1990.

4.2.7.2 Powys County Council

Box 4.5 *Powys County Council v Mr. E.J. Price* [2017] EWCA Civ 1133

A landfill had operated on land, owned by Mr Price and Ms Hardwicke, until 1992. Up until the Transco case (see Box 4.4), Powys County Council (PCC) had believed that it was responsible for any contamination caused by its predecessors. PCC had therefore monitored the site and operated a filtration and treatment plant, with a pumping station for leachate, since 2001. However, following the Transco case, PCC asserted that it was not liable for the landfill operated by its predecessor.

Mr Price and Ms Hardwicke were concerned that their land might be identified as contaminated land, and therefore sought a declaration that PCC was liable as an 'appropriate person'. The Court of Appeal in this case followed Trancso, and held that PCC was not liable for the past conduct of its predecessor. Moreover, the predecessor only operated the landfill until 1992, while PCC was not created as a successor until 1996, and part IIA of the Environmental Protection Act 1990 did not come into force in Wales until 2001.

4.2.7.3 Statutory transfer liability

It should be noted that the decisions in the National Grid Gas plc and Powys County Council cases do not preclude liability for statutory predecessors, as the judgements depended on the framing of the transfer legislation.

4.2.8 Critique of the contaminated land regime

The contaminated land regime has been criticised because developers may be discouraged from developing contaminated land, due to the risk of introducing pathways and receptors during construction. In addition, local authorities are concerned about the expense of possible appeals to remediation notices, including legal fees and funding of the remediation costs themselves (Brown, 2016).

4.2.8.1 Unworkable regime

Fogleman (2014b) argued that part IIA of Environmental Protection Act 1990 is unfit for purpose as primary delegation of its implementation and enforcement to local authorities has resulted in an unworkable regime that has failed to meet its objectives. She suggested transferring responsibility to the Environment Agency and imposing modified joint and several liability as default for all persons who caused or knowingly permitted contamination of land (Fogelman, 2014a).

Lees (2016) identified five models of liability for remediating contaminated land, with France, Spain and Sweden having adopted the same model as the UK.

4.2.8.2 Finance, land use and site investigations

A research group discussion on contaminated land focused on three main issues: finance, land use and site investigations. It was argued that it is difficult to value contaminated land due to the uncertainty of grant provision. It was recommended that the ultimate use of such land and the consequent necessity of remediation should be scrutinised. Site investigations were advocated to market contaminated land, but the expense of comprehensive investigations was recognised (Charlson, 2018).

4.3. Environmental Damage (Prevention and Remediation) (England) Regulations 2015

The Environmental Damage (Prevention and Remediation) (England) Regulations 2015 (the Environmental Damage Regulations) apply to the most serious cases of environmental damage. They relate to the prevention and remediation of damage to species, habitats, sites of special scientific interest, surface or groundwater, and land.

4.3.1 Operator

Under the Environmental Damage Regulations, an operator must

- take all practical steps to prevent environmental damage
- notify the relevant regulator of imminent threats of environmental damage
- notify the relevant regulator of an activity that has caused environmental damage
- provide information to regulators on request, to enable them to perform their duties.

If environmental damage has already occurred, the operator must take all practicable steps to prevent further damage.

4.3.2 Enforcement

If the regulator considers that environmental damage has occurred, it can serve a remediation notice on the responsible operator, setting out the measures that must be taken. The principal enforcement authority is the relevant local authority.

Failure to comply with a remediation notice, without reasonable excuse, is a criminal offence punishable by a fine. The regulator can carry out the remediation itself and recover the costs from the relevant parties.

4.4. Brownfield land

Brownfield land is defined in the National Planning Policy Framework.

The Town and Country Planning (Brownfield Land Register) Regulations 2017 mandated LPAs to prepare, maintain and publish a brownfield land register. Charlson (2021) examined the introduction of brownfield land registers in England.

4.4.1 Brownfield land definition

Brownfield land is defined in the National Planning Policy Framework (HCLG, 2021) as

> Land which is or was occupied by a permanent structure, including the curtilage of the developed land (although it should not be assumed that the whole of the curtilage should be developed) and any associated fixed surface infrastructure. This excludes:
>
> - land that is or was last occupied by agricultural or forestry buildings
> - land that has been developed for minerals extraction or waste disposal by landfill, where provision for restoration has been made through development management procedures
> - land in built-up areas such as residential gardens, parks, recreation grounds and allotments, and
> - land that was previously developed but where the remains of the permanent structure or fixed surface structure have blended into the landscape.

4.4.2 Permission in principle for development of land

Section 150 of the Housing and Planning Act 2016 concerns 'permission in principle for development of land'. The section contains provisions to grant permission in principle for housing-led development of land in England.

This section inserted a new section into the Town and Country Planning Act 1990, giving the Secretary of State the power, by development order, to grant planning permission in principle to land that is allocated for development in a 'qualifying document'.

4.4.3 Local planning authority registers

Section 151 of the Housing and Planning Act 2016 requires LPAs to keep registers of particular kinds of land. The section authorised the Secretary of State to make regulations requiring an LPA in England to prepare, maintain and publish a register of land within the authority's area which is of a prescribed description or satisfies prescribed criteria.

4.4.4 Permission in Principle Regulations and Order

Two statutory instruments were subsequently made: the Housing and Planning Act 2016 (Permission in Principle etc.) (Miscellaneous Amendments) (England) Regulations 2017 and the Town and Country Planning (Permission in Principle) Order 2017.

4.4.5 Brownfield land registers

However, neither section 150 nor section 151 of the Housing and Planning Act 2016 specifically refer to brownfield land. Therefore, secondary legislation and statutory guidance were required for brownfield land.

The Town and Country Planning (Brownfield Land Register) Regulations 2017 (the Brownfield Regulations) were made by the Secretary of State for Communities and Local Government in exercise of powers conferred by sections 14A and 122(1)(a) and 122(3) of the Planning and Compulsory Purchase Act 2004.

Regulation 3 of the Brownfield Regulations mandated LPAs to prepare, maintain and publish brownfield land registers. Regulations 4 and 5 require land, where it meets the following criteria, to be entered in part 1 of the register.

- The land is at least 0.25 hectares or can support at least five dwellings.
- The land is suitable for residential development, and
  - is allocated in the local plan
  - has planning permission
  - has permission in principle, or
  - is appropriate for residential development.
- 3. The land is available for residential development, and
  - owner has expressed an intention to sell or develop the land
  - no ownership issues or other legal impediments.

- 4. Residential development of the land is achievable.
  - development likely within 15 years of the entry date.

4.4.6 Permission in principle guidance

The Department for Levelling Up, Housing and Communities guidance on permission in principle (DLUHC, 2019, Paragraph: 001 Reference ID: 58-001-20180615) explains that

> The permission in principle consent route is an alternative way of obtaining planning permission for housing-led development which separates the consideration of matters of principle for proposed development from the technical detail of the development. The permission in principle consent route has two stages: the first stage (or permission in principle stage) establishes whether a site is suitable in principle and the second ('technical details consent') stage is when the detailed development proposals are assessed.

The guidance details that LPAs can grant permission in principle to a site upon receipt of a valid application or by entering a site in part 2 of its brownfield land register, which will trigger a grant of permission in principle for that land providing the statutory requirements set out in the Town and Country Planning (Permission in Principle) Order 2017 (as amended) and the Brownfield Regulations are met. The requirements for a valid permission in principle application are set out in Article 5D of the Town and Country Planning (Permission in Principle) Order 2017 (as amended). The scope of permission in principle is limited to location, land use and amount of development.

4.4.7 Publication of site information

Schedule 2 of the Brownfield Regulations sets out the information that an LPA is required to publish about a site, including the location and the type and amount of development. The amount of development must be expressed as a range, indicating the minimum and maximum net number of dwellings (i.e. taking into account any existing dwellings on the site) that are, in principle, permitted.

4.4.8 Consultation

The consultation that must be undertaken before sites can be granted permission in principle is set out in the Brownfield Regulations for suitable sites on brownfield land registers, and in the Town and Country Planning (Permission in Principle) Order 2017 (as amended) when permission in principle is sought by application. In both situations, the LPA must consult bodies identified in schedule 4 to the Town and Country Planning (Development Management Procedure) (England) Order 2015 if, in their opinion, the land within the site falls within the prescribed category, and take into account any responses received.

4.4.9 Review suitability of all relevant sites

LPAs should, in accordance with the permission in principle guidance, regularly review the suitability of all relevant sites on their brownfield land registers for a grant of permission in principle, taking into account relevant policies in the development plan. In addition, relevant existing information sources include historic environment records and advice published by statutory consultees.

4.4.10 No right of appeal

The permission in principle guidance explains that there is no right of appeal where an LPA decides not to enter a site in part 2 of a brownfield land register and trigger the grant of permission in principle. Furthermore, there is no power for the Secretary of State to call in a decision on whether to enter a site in part 2 of a brownfield land register. Site allocations in existing local or neighbourhood plans do not have a grant of permission in principle.

4.4.11 Publicising sites

Regulation 6 of the Brownfield Regulations and Article 5G of the Town and Country Planning (Permission in Principle) Order 2017 (as amended) set out the statutory requirements for publicising sites entered on brownfield land registers and when a valid application has been received. Site and online notices are required. Furthermore, LPAs must publicise their intention to grant sites permission in principle with site and online notices.

4.4.12 Technical details consent

The permission in principle guidance states that, following a grant of permission in principle, the site must receive a grant of technical details consent for a grant of planning permission for the development. Technical details consent can be obtained by submission of a valid application, in accordance with the permission in principle, to the LPA. The requirements for a valid technical details consent application are the same as those for an application for full planning permission. The applicant must also complete an ownership certificate, which confirms that an appropriate notice has been served on landowners. LPAs may agree to planning obligations at this stage and the Community Infrastructure Levy may also apply.

The publicity requirements for technical details consent applications mirror the approach taken for planning applications. Article 15 of the Town and Country Planning (Development Management Procedure) (England) Order 2015 sets out the minimum statutory requirements.

4.4.13 Review of brownfield land registers

The Autumn Budget and Spending Review 2021 included £1.8 billion to bring 1500 hectares of brownfield land into use for housing (HM Treasury, 2021).

The countryside charity CPRE reports annually on the status and potential of brownfield land in England. Its '*State of Brownfield Report 2022* (CPRE, 2022) reviewed local councils' registers of brownfield land and found that over 1.2 million homes could be built on 23 000 sites covering more than 27 000 hectares of previously developed land. Just 45% of available housing units have been granted planning permission and 550 000 homes with planning permission are still awaiting development.

References

Statutes

Environmental Protection Act 1990
Housing and Planning Act 2016
Planning and Compulsory Purchase Act 2004
The Town and Country Planning Act 1990

Regulations and orders

The Contaminated Land (England) Regulations 2006
The Environmental Damage (Prevention and Remediation) (England) Regulations 2015
The Housing and Planning Act 2016 (Permission in Principle etc.) (Miscellaneous Amendments) (England) Regulations 2017
The Town and Country Planning (Brownfield Land Register) Regulations 2017
The Town and Country Planning (Development Management Procedure) (England) Order 2015
The Town and Country Planning (Permission in Principle) Order 2017

Case law

Circular Facilities (London) Ltd v Sevenoaks District Council [2005] EWGC 865 (Admin)
Crest Nicholson Residential Ltd, R (On the Application Of) v Secretary of State for Environment, Food and Rural Affairs & Ors [2010] EWHC 561 (Admin)
Powys County Council v Mr. E.J. Price & Anor [2017] EWCA Civ 1133
R (On the Application of National Grid Gas plc (formerly Transco plc)) v Environment Agency [2007] UKHL 30
Redland Minerals Ltd, R (On the Application Of) v Secretary of State for Environment, Food and Rural Affairs [2010] EWHC 913 (Admin)

Journals

Brown L (2016) The contaminated land regime and austerity. *International Journal of Law in the Built Environment* **8(3)**: 210–225.

Charlson J (2018) Regeneration of brownfield land: the environmental law challenges. *Journal of Property, Planning and Environmental Law* **10(3)**: 202–218.

Charlson J (2021) The introduction of brownfield land registers in England. *Planning Practice & Research* **36(2)**: 216–229.

Fogleman V (2014a) The contaminated land regime: time for a regime that is fit for purpose (part 2). *International Journal of Law in the Built Environment* **6(1/2)**: 129–151.

Fogleman V (2014b) The contaminated land regime: time for a regime that is fit for purpose (part 1). *International Journal of Law in the Built Environment* **6(1/2)**: 43–68.

Lees E (2016) The polluter pays principle and the remediation of land. *International Journal of Law in the Built Environment* **8(1)**: 2–20.

Websites

CPRE (2022) *State of Brownfield Report 2022*. https://www.cpre.org.uk/resources/state-of-brownfield-report-2022 (accessed 09/03/2023).

Defra (Department for Environment, Food and Rural Affairs) (2012) *Environmental Protection Act 1990: Part 2A. Contaminated Land Statutory Guidance*. https://assets.publishing.service.gov.uk/government/uploads/system/uploads/attachment_data/file/223705/pb13735cont-land-guidance.pdf#:~:text=The%20contaminated%20land%20regime%20under%20Part%202A%20of,the%20environment%2C%20where%20there%20is%20no%20alternative%20solution (accessed 09/03/2023).

DLUHC (Department for Levelling Up, Housing and Communities) (2019) Permission in principle. https://www.gov.uk/guidance/permission-in-principle (accessed 09/03/2023).

Environment Agency (2021) Land contamination risk management (LCRM). https://www.gov.uk/government/publications/land-contamination-risk-management-lcrm (accessed 09/03/2023).

HCLG (Ministry of Housing, Communities and Local Government (2021) National Planning Policy Framework. https://www.gov.uk/government/publications/national-planning-policy-framework–2 (accessed 09/03/2023).

HM Treasury (2021) *Autumn Budget and Spending Review 2021*. https://www.gov.uk/government/publications/autumn-budget-and-spending-review-2021-documents (accessed 09/03/2023).

Baker F and Charlson J
ISBN 978-0-7277-6645-8
https://doi.org/10.1680/elsc.66458.119

Chapter 5
Waste management

Francine Baker

5.1. Introduction

This chapter concerns the main strategy, plans and legislation concerning waste management for the construction/engineering industry. 'Construction and demolition waste' is given a wide definition under the law. It means waste from the preparatory work, and from any improvement, repair or alteration of the construction or demolition works. It includes any items that had been part of an infrastructure. It also includes waste from the exploration or extraction of mineral resources. Whether it is builders generating waste or contractors dealing with waste (such as conducting groundworks) or operators of sites where the waste ends up, they all have legal obligations to classify the waste produced and code it correctly, and make sure that it goes to the right place. Common examples of construction and demolition waste include baths, sinks, toilets, doors, fences, paths, concrete posts, rubble, walls, brickwork, pipework, wiring, plasterboard, radiators, sinks and windows (Environment Agency, 2022a).

There has been considerable change in waste management planning, policy and regulation in the past 20 years. Following Brexit, from 31 December 2020, movements of waste between the UK and the European Union (EU) are subject directly to the Basel Convention on the Control of Transboundary Movements of Hazardous Wastes and Their Disposal. EU regulation on the shipment of waste (Regulation (EC) No 1013/2006) is now defunct. However, most of the UK's existing waste legislation is a result of EU law, and much law has been kept as part of UK law through the European Union (Withdrawal) Act 2018. On 11 March 2020, as part of its European Green Deal, the European Commission adopted a new Circular Economy Action Plan (CEAP). In July 2020, the Department for Environment, Food and Rural Affairs (Defra, in England), the Scottish Government, the Welsh Government and the Department of Agriculture, Environment and Rural Affairs (Northern Ireland) jointly issued the Circular Economy Package (CEP) policy statement (Defra, 2020a), which sets out the key changes made by the CEAP and the approach of the UK to the transposition of the 2020 CEAP measures.

EU directives may, therefore, still be relevant for the purpose of understanding and interpreting certain provisions of retained EU-derived domestic legislation. For example, the Waste (Circular Economy) (Amendment) Regulations 2020, which came into force on 1 October 2020, transposed into UK law six amending EU directives concerning waste, including the 2020 CEP measures for England and Wales. Therefore, recent approaches to waste management are focused on the move to a circular economy. Overall, more waste is being reused and recycled, and less goes to landfill.

5.1.1 Chapter contents summary

This chapter covers the legal definition of waste, the end of waste test, the waste hierarchy and duty of care, and the scope of the residual waste target. It describes how to reduce waste, the duties of waste collectors, the transport and handling of waste, compliance requirements, as well as the misdescription of waste and landfill. The various reforms under the Environment Act 2021 should make a significant contribution to waste management when the relevant regulations come into force. The reforms include charging schemes, the setting of waste management targets, mandatory electronic waste tracking, updated duties regarding the separation of recyclates, and the duties of waste collectors, as well as compliance requirements. Waste and planning considerations, and the legislation governing the environmental permitting regime for dealing with waste are discussed in Chapter 2. The current regime for dealing with hazardous waste until regulations are operating under the Environment Act 2021 is dealt with in Section 7.5 in Chapter 7.

This chapter does not include discussion of legislation covering some specialist areas that are relevant to waste management. For example, it does not discuss legislation relating to the Waste Electrical and Electronic Equipment Directive (recast) (WEEE Directive, 2012/19/EU), the Battery Directive (2006/66/EC) or the End of Life Vehicles Directive (2000/53/EC), or Regulation (EC) No 1013/2006 on the shipment of waste.

A financial incentive, in the form of the landfill tax, for businesses to divert commercial and industrial waste away from landfill has led to less landfill and increased public and private investment in waste management facilities. This chapter will therefore explain the landfill tax regime, exemptions from the tax and recent developments.

5.2. Strategy, plans and regulation

The UK's 25-year environment plan, published in 2018 (Defra, 2021a; HMG, 2018), set out the government's resources and waste management strategy. There is also the 2018 resources and waste strategy for England (Defra and EA, 2018) and the 2010 waste strategy for Wales (Welsh Government, 2010). One of the main goals in the UK government's 25-year plan is to eliminate waste of all kinds by 2050. Part of the plan is to use resources from nature more sustainably and efficiently, to produce related policies to increase resource efficiency, and to reduce pollution and waste (HMG, 2018, p. 10). The goal is for these policies and plans to help increase the amount of waste that is recovered or recycled (Defra, 2022a). The government's first revision of the 25-year plan, the Environmental Improvement Plan 2023, addresses recent legislative developments (Defra, 2023).

5.2.1 Resources and waste strategy 2018

This strategy (Defra and EA, 2018) is concerned with preventing and preserving resources by minimising waste. It identifies certain materials in the construction and demolition sector as a priority area. However, when published in 2018, the scope of materials was yet to be defined (Defra and EA, 2018, p. 39), although the strategy states that it will work with the construction industry to produce such a scope. It also states that it will work with the Green Construction Council to reduce waste and increase resource efficiency through the adoption of circular economy principles, and to develop a road map to achieve zero avoidable waste (Defra and EA, 2018, pp. 45–46).

The strategy advocates more sustainable government procurement to help to generate less waste, the use of building information modelling (BIM) to help avoid inefficiencies such as a lack of coordination in the design of major construction projects, and to build new infrastructure in a 'modular' way, which the strategy favours (Defra and EA, 2018, pp. 63–64).

5.2.2 Waste management plan for England 2021

This revised plan (Defra, 2021b), pursuant to the Waste (England and Wales) Regulations 2011 (as amended), focuses on how to manage waste, and is produced and reviewed every six years. The first plan was produced in 2013. It is a non-site-specific document that provides an analysis of waste management in England. It is supplemented by a waste prevention programme which is concerned with how to prevent materials becoming waste through reuse, repair and remanufacture. The statutory waste management plan for Wales is set out in the document *Towards Zero Waste* (Welsh Government, 2010), and there is an associated suite of sector plans and other documents.

The waste management plan and programme should be read in conjunction with the government's national planning policy for waste (DLUHC, 2014), which is updated periodically. This sets out the framework within which local authorities can produce strategies to identify, through their local plans, any suitable sites for facilities to address waste management needs in their area.

Waste planning authorities (WPAs) are authorities with the responsibility for the use of the land in terms of the control of waste management. WPAs are either county councils or unitary authorities, including National Park authorities.

A WPA is responsible for producing a waste development framework for its area. This is a set of local policy documents that determines the type of waste management facilities and where they are located. The framework also establishes appropriate criteria on which to base planning decisions. WPAs are also responsible for determining planning applications, and controlling waste management facilities through the planning system.

5.2.3 Waste (Miscellaneous Amendments) (EU Exit) (No. 2) Regulations 2019

These Waste (Miscellaneous Amendments) (EU Exit) (No. 2) Regulations 2019 (the EU Exit Regulations) amended domestic legislation, including the waste transport regulations and the Waste (England and Wales) Regulations 2011, which have since been amended. A feature of the EU Exit Regulations is the waste hierarchy, which had been required by Article 4(1) of the Waste Framework Directive (2008/98/EC) to be applied by the EU member states in their waste legislation and policy. On 11 March 2020, as part of its European Green Deal, the European Commission adopted a new Circular Economy Action Plan (CEAP).

Subsequently, in July 2020, Defra (in England), the Scottish Government, the Welsh Government and the Department of Agriculture, Environment and Rural Affairs (Northern Ireland) jointly issued the Circular Economy Package (CEP) policy statement (Defra, 2020a), which sets out the key changes made by the CEAP and the approach of the UK to transposition of the 2020 CEAP measures. The Waste (Circular Economy) (Amendment) Regulations 2020, which came into force on 1 October 2020, made the legislative changes required to transpose the 2020 CEP measures on behalf of England and Wales.

5.2.4 Environment Act 2021

The UK Environment Act 2021 provides a new framework for environmental protection. The Act is a vehicle for Defra's environmental policies. It sets out the legal framework for significant reforms to local authority waste and recycling services, and establishes a new relationship between central and local government regarding environmental improvement. The Act amends existing legislation, including the Environmental Protection Act 1990, regarding dealing with the separation of waste, electronic waste tracking, hazardous waste and transfrontier shipments of waste. The Act also amends the Environment Act 1995 by granting powers to establish charging schemes.

The government has placed waste management responsibility on the producer. This also assists an environmentally aware management of waste.

5.3. What is waste?

The legal definition of waste is simply 'any substance or object which the holder discards or intends or is required to discard' (Defra, 2021c). The definition comes from Article 1(3) of the Waste Framework Directive (2008/98/EC) as amended by the Waste (Circular Economy) (Amendment) Regulations 2020.

However, the meaning of the words in the above definition (e.g. the words 'substance', 'object' and 'discard') are subject to interpretation in the courts, which is found in case law.

It also should be considered whether or not a substance or an object resulting from a production process is waste or a by-product of that process, as per Article 5 (1). Again, judges' interpretations in case law may need to be considered. However, you can find out if your material meets the status of a by-product by undertaking an assessment yourself.

The Environment Agency for England provides detailed advice about whether a material is waste, is a by-product or meets 'end of waste' status on its webpages (Environment Agency, 2022b).

5.3.1 End of waste test

What is legally considered to be a waste material may become non-waste, and then the regulations will not apply to that material (substance or object), regardless of whether the substance has been imported or exported. An example could be a material that has been recycled or recovered in some way, for example, through a simple or highly complex process that has removed contaminants or changed the material. For a material to no longer be considered waste, it must satisfy the end of waste test in Articles 6(1) and 6(2) of the Waste (Circular Economy) (Amendment) Regulations 2020.

However, if the material is iron, steel or aluminium scrap, glass cullet or copper scrap, you will also need to meet the requirements of other regulations that the UK kept following Brexit. These are: Council Regulation (EU) No. 333/2011 (iron, steel or aluminium scrap), Commission Regulation (EU) No. 1179/2012 (glass cullet) and Commission Regulation (EU) No. 715/2013 (copper scrap).

If, however, there is an inability to follow a relevant quality protocol produced by the Environment Agency, the material will not be classed as waste. If there is no quality protocol for your material, then the detailed guidance from the Environment Agency (2021a) can be used to explain how you will assess the waste status of the material and whether you are planning to produce a new product from waste. The guidance also explains when waste rules apply to your material, import and export implications, and the UK environmental regulators.

5.3.2 Self-assessment advice or Environment Agency opinion

If you decide to assess whether your material meets the status of a by-product by undertaking an assessment yourself, any guidance issued by the Environment Agency and by Defra must be followed. However, it is still possible that the Environment Agency may disagree with your self-assessment.

Alternatively, an opinion may be obtained from the Environment Agency (Environment Agency, 2021b). At the time of writing, an opinion from the Environment Agency costs £125 per hour.

Guidance on how to check if your material is waste is available on the government website (Environment Agency, 2022b).

5.4. Waste hierarchy and duty of care

Defra's detailed evidence report *Resource Efficiency and Waste Reduction Targets* (Defra, 2022a) refers to the waste-hierarchy structure in the Waste Regulations (England and Wales) 2011 (regulation 12). The regulations place a duty on anyone who imports, produces, collects, transports, recovers or disposes of waste, and to any dealer or broker who has control over waste or is involved in waste to take reasonable steps to apply this waste hierarchy when they transfer waste. Within the hierarchy, the prevention of waste is the ideal position to achieve, and disposal of waste is the least desirable. The hierarchy is

1. prevention
2. preparing for reuse
3. recycling
4. other recovery
5. disposal.

Defra's report *Resource Efficiency and Waste Reduction Targets* (Defra, 2022a) confirms the government's goal for waste to be driven further up the waste hierarchy and to be retained within a circular economy.

Section 34(1A) of the Environmental Protection Act 1990 states that

> It shall be the duty of any person who imports, produces, carries, keeps, treats or disposes of controlled waste or, as a broker, has control of such waste, to take all such measures applicable to him in that capacity as are reasonable in the circumstances ...

To comply with the duty of care, a waste holder should consider what the courts may require as evidence of compliance, and act on that basis. This will often require audits of others in the waste chain (e.g. producer, carrier or treater) as part of demonstrating compliance with this section of the Act. It will also usually require evidence showing that those managing wastes at waste management facilities such as transfer stations

1. have a permit that allows them to manage the waste in question, and comply with any permit conditions
2. ensure that all holders take measures to prevent the escape of waste
3. ensure that wastes are transferred either to a registered waste carrier or permit holder
4. provide relevant written information upon waste transfer, in accordance with regulation 35 of the Waste (England and Wales) Regulations 2011.

Waste producers bear responsibility for ensuring legally compliant waste management, and this responsibility cannot be devolved to others in the waste chain. Regulators have the power to enforce the 'duty of care', but they do not have a duty to do so.

The case described in Box 5.1 demonstrates that waste producers must act in a way that is 'reasonable in the circumstances'.

Box 5.1 *Mountpace Ltd v London Borough of Haringey* [2012] EWHC 698 (Admin)

A business (Mountpace Ltd) had commissioned a developer to renovate a property, and this work required the removal and disposal of waste from the site.

The developer subcontracted the removal and disposal of waste to a third-party waste contractor, and this contractor illegally dumped (fly-tipped) the waste. Mountpace Ltd was charged with failure to comply with section 34 of the Environmental Protection Act 1990, but argued that it was not reasonably foreseeable that the developer would subcontract the work to an unauthorised contractor who would fly-tip the waste. The High Court ruled that the Mountpace Ltd was in breach of the law because of its failure to exercise control over what was happening to waste on its premises. The judge stated:

> What is important … was … that there would be transfers of controlled waste from the premises yet [*the business*] had chosen to distance itself from the site and not to preserve a permanent presence on the site and not to play any direct role in the transfer of the waste'.

5.5. Scope of the residual waste target

The resource efficiency and waste reduction target (REWDT) is to reduce residual waste (Defra, 2022a). This target is consistent with all government policies and plans and the Environment Act 2021. Residual waste is waste that is not reused or recycled. Reducing this waste would mean reducing the amount of waste that is sent elsewhere (e.g. to landfill or incineration). However, the scope of the REWDT does not include what it classes as major mineral wastes, using Eurostat definitions (Eurostat, 2023). It therefore does not include construction and demolition waste (e.g. concrete, bricks and sand), soils or mineral wastes from excavation and mining activities. However, it does capture biodegradable materials from construction and demolition, such as wood waste. The reason given for the REWDT excluding major mineral wastes from construction, demolition and excavation (CDE) is that the date regarding CDE waste is uncertain, and therefore a meaningful long-term target cannot be set (Defra, 2022a, pp. 20–23).

However, the Defra report *Resource Efficiency and Waste Reduction Targets* (Defra, 2022a) refers to the waste prevention programme as playing a role in reducing CDE waste, as does the Environment Act 2021, which allows for targets to be set in the future for reducing other CDE residual waste or mineral waste. The Act also introduces an electronic waste tracking service that will enable more efficient and accurate collection of data on the movements of waste. This will enable appropriate targets for CDE waste to be set in the future (Defra, 2022a, p. 11).

The Environmental Targets (Residual Waste) (England) Regulations 2023 made under the Environment Act 2021 section 1 came into force on 30 January 2023. Regulation 2 sets a long-term target – the total mass of residual waste per head of population in England does not exceed 287 kilograms by 31 December 2042.

Residual waste routes include those generated by the construction industry i.e.,

(*a*) sent to landfill
(*b*) put through incineration into the United Kingdom (UK)
(*c*) use in energy recovery (UK)
(*d*) sent outside the UK for energy.

The meaning of ‘energy recovery’ includes any waste treatment, excluding anaerobic digestion, which generates energy such as electricity or heat or which converts the waste into other energy products such as fuels and substitute natural gas.

The Environmental Improvement Plan 2023 states that (the government) ‘... will halve ‘residual’ waste (excluding major mineral waste) produced per person by 2042. This is underpinned by a number of interim targets, by 31 January 2028’ (Defra, 2023).

5.5.1 Commercial and industrial waste statistics

According to Defra official statistics (Defra, 2022b), CDE (including dredging) generated around three-fifths (62%) of total UK waste in 2018, commercial and industrial (CI) waste accounted for almost a fifth (19%) and the remaining fifth was split between ‘households’ (12%) and ‘other’ activities (7%). In England, the share of CDE waste was higher, at 64% of the total.

The UK generated an estimated 43.9 million tonnes of CI waste in 2018. Of this, 37.2 million tonnes (85%) was generated in England. Estimates for 2020 indicate that CI waste generation was around 33.8 million tonnes in England, which is a decrease from 2018 of 2.4 tonnes (Defra, 2022b).

5.5.2 How to reduce waste

The Defra report *Resource Efficiency and Waste Reduction Targets* recommends that CDE waste can be reduced by (Defra, 2022a, p. 24)

- using fewer materials (e.g. by extending the life of a building in use or by design)
- prevention of waste (e.g. by using lean production techniques during construction)
- selecting materials and building techniques that enable recycling and reuse of materials at the end of life.

The recent Zero Avoidable Packaging (ZAP) Waste in Construction project (Adams, 2022) highlights that the construction industry needs to do more to reduce plastic packaging waste on construction sites. It recommends that the industry implement a packaging hierarchy order as follows

- elimination by removing packaging altogether (e.g. use of bulk deliveries)
- reduction (e.g. use of larger pack sizes), optimisation (e.g. lighter weight)
- reuse (e.g. reusable crates)
- recycling into new products
- recovery (energy from waste)
- disposal to landfill.

5.6. Electronic tracking reforms

Part 3, section 58 of the Environment Act 2021 allows for the government to introduce mandatory electronic waste tracking in the UK. These provisions came into force on 9 January 2022. This part of the Act fulfils a commitment made in the 2018 resources and waste strategy for England (Defra and EA, 2018), as it is intended to provide a comprehensive way to see what is happening to the waste produced in the UK. This will help to support more effective regulation of waste and help businesses to comply with their duty of care with regard to waste. In addition, it will assist in moving towards a more circular economy, and help prevent waste and detect and deter waste crime. It is expected that the digital waste tracking service will be in operation by 2024. It should enable a single central source of up-to-date information on where and how waste is created, who is handling it, what is done to it and where it ends up.

The consultation document on mandatory digital waste tracking (Defra, 2022d) proposed that the tracking service will include controlled waste (both hazardous and non-hazardous household, commercial and industrial waste) and extractive waste (from mines and quarries). However, the details of non-hazardous waste that is treated, discharged, disposed of or recovered at the site of production will not need to be recorded. Various enforcement options were for failure to record information, intentionally or recklessly providing incomplete or false information in a digital record, and moving or receiving waste without a digital record.

In April 2022, Defra published a policy paper (Defra, 2022c) which confirms that the government will also make legislative amendments regarding the waste duty of care, hazardous waste, transfrontier shipments of waste, waste permitting and licensing, and the statutory use of the waste data flow system by local authorities in Wales. It notes that the waste duty of care codes of practice will also need to be revised.

5.7. Duties of waste collectors – to keep different recyclates separate

Section 57 of the Environment Act 2021 amends sections 45 and 45A of the Environmental Protection Act 1990 by adding many sections concerning obligations regarding the recycling of material, (e.g. glass, metal, plastic, paper and card, and food waste), the duties of waste collectors and compliance matters.

A new section, 45AZB, of the Environmental Protection Act 1990 concerns the separate collection of industrial or commercial waste in England, and section 45AZD concerns the duties of waste collectors of the latter.

> Section 45AZB–England: separate collection of industrial or commercial waste
>
> (1) This section applies in relation to arrangements for industrial or commercial waste to be collected from premises in England by a person who, in collecting the waste
> (*a*) is acting in the course of a business (whether or not for profit), or
> (*b*) is exercising a public function (including a function under section 45(1)(b) or (2)).
>
> (2) So far as they relate to waste which is similar in nature and composition to household waste ('relevant waste' (glass; metal; plastic; paper and card; and food waste)) the arrangements must meet the conditions in subsections (3) to (7). This is subject to any provision in regulations under section 45AZC (any exemptions the Secretary of State may make).
>
> (3) The first condition is that recyclable relevant waste must be collected separately from other relevant waste.
>
> (4) The second condition is that recyclable relevant waste must be collected for recycling or composting.
>
> (5) The third condition is that recyclable relevant waste in each recyclable waste stream must be collected separately, except so far as provided by subsection (6).
>
> (6) Recyclable relevant waste in two or more recyclable waste streams may be collected together where
> (*a*) it is not technically or economically practicable to collect recyclable relevant waste in those recyclable waste streams separately, or
> (*b*) collecting recyclable relevant waste in those recyclable waste streams separately has no significant environmental benefit (having regard to the overall environmental impact of collecting it separately and of collecting it together).
>
> (7) But recyclable relevant waste within subsection (10)(a) to (d) may not be collected together with recyclable relevant waste within subsection (10)(e).
>
> (8) The person who presents relevant waste from the premises for collection under the arrangements must present it separated in accordance with the arrangements.
>
> (9) This subsection does not apply so far as the person is subject to an equivalent duty by virtue of a notice under section 47.
>
> (10) Relevant waste is 'recyclable relevant waste' if
> (*a*) it is within any of the recyclable waste streams, and
> (*b*) it is of a description specified in regulations made by the Secretary of State.

(11) For the purposes of this section the 'recyclable waste streams' are
   (*a*) glass
   (*b*) metal
   (*c*) plastic
   (*d*) paper and card
   (*e*) food waste.

5.7.1 Who is a waste collector

The new section, 45AZD, of the Environmental Protection Act 1990 states that

Section 45AZD–sections 45A to 45AZB: duties of waste collectors

(1) Subsection (2) applies where
   (*a*) a person collects or proposes to collect waste under arrangements to which section 45A, 45AZA or 45AZB (industrial or commercial waste) applies, and
   (*b*) the arrangements include arrangements to collect recyclable household waste or recyclable relevant waste in two or more recyclable waste streams together in reliance on sections 45A(6), 45AZA(6) or 45AZB(6).

Section 47(3) of the Act deals with receptacles for commercial or industrial waste. It has been amended so that an English waste collection authority may require separate receptacles or compartments of receptacles to be used for the purposes of complying with section 45AZB.

5.7.2 Compliance notices

Under the subsection (1) of the new section 45AZF of the Environmental Protection Act 1990, compliance notices may be issued where a person other than an 'English waste collection authority' (i.e. a local authority) fails to comply with section 45AZB (the separate collection of industrial or commercial waste in England) (Environment Agency, 2016).

The following subsections state what the notice should contain and the consequences of failure to comply with the notice.

Section 45AZF–sections 45AZA and 45AZB: compliance notices

(2) It [*the Environment Agency*] may give that person a notice (a 'compliance notice') requiring them to take specified steps within a specified period to secure that the failure does not continue or recur.
(3) A compliance notice must
   (*a*) specify the failures to comply with section 45AZA or 45AZB
   (*b*) specify the steps which must be taken for the purpose of preventing the failure continuing or recurring
   (*c*) specify the period within which those steps must be taken, and
   (*d*) give information as to the rights of appeal (including the period within which an appeal must be brought).
(4) A person who fails to comply with a compliance notice commits an offence.
(5) A person who commits an offence under subsection (4) is liable on summary conviction or conviction on indictment to a fine.

5.7.3 Compliance notice appeals

The new section 45AZG of the Environmental Protection Act 1990 concerns appeals against compliance notices.

Section 45AZG–sections 45AZA and 45AZB: appeals against compliance notices

(1) A person who is given a compliance notice may appeal to the First Tier Tribunal against
  (*a*) the notice, or
  (*b*) any requirement in the notice.
(2) The notice or requirement has effect pending the determination of the appeal unless the tribunal decides otherwise.
(3) The tribunal may
  (*a*) quash the notice or requirement
  (*b*) confirm the notice or requirement
  (*c*) vary the notice or requirement
  (*d*) take any steps the Environment Agency could take in relation to the failure giving rise to the notice or requirement, or
  (*e*) remit any matter relating to the notice or requirement to the Environment Agency.

5.8. Hazardous waste

Under the Hazardous Waste (England and Wales) Regulations 2005, waste producers must classify their waste, and store hazardous waste separately from their commercial waste. All waste must be identified, assessed and classified before disposal or recycling.

Waste producers must ensure that a licensed carrier collects their waste by working with a registered company and checking that the appropriate environmental permits are in place before engaging the collection.

5.8.1 Consignment note

Every time waste is collected, a consignment form must be filled out. Each part of the form must be completed by the appropriate person at the right time. Both the carrier and the producer must keep a copy of this note. At least three copies must be kept.

A consignment note form is downloadable and details of how to complete the note are given on the government website (HMG, 2023a).

5.8.2 The mixing of hazardous waste

The mixing of hazardous waste is prohibited. The following waste must not be mixed

- a hazardous waste with non-hazardous waste
- a hazardous waste with a non-waste
- different types (categories) of hazardous waste with each other
- waste oils with different characteristics.

Any mixing of hazardous waste must have been specifically authorised by an environmental permit (see Section 2.12 in Chapter 2) and the best available technique must be used.

5.8.3 Hazardous waste – four categories

- Universal wastes – batteries, or equipment containing mercury.
- Mixed wasted – deemed radioactive, or containing hazardous waste components.
- Characteristic wastes – corrosive, toxic or reactive waste.
- Listed wastes, as determined by the Environmental Protection Agency (EPA, 2023) – wastes from the F (waste from non-specific sources) and K (source-specific waste) lists.

Guidance on how to classify waste and how to deal with or dispose of it is given on the government website (HMG, 2023b).

5.8.4 Hazardous waste characteristics

Typical characteristics of hazardous waste include the following.

- *Potentially flammable*. Materials are flammable when they reach a specific temperature. For example, spills and rags that are contaminated with cleaning products, oils or paint, and oil and fuel filters are potentially flammable.
- *Ignitable*. Materials can potentially cause a fire during storage, transport or disposal. For example, fluorescent tubes and sodium lamps (because they contain sodium and alkali metals).
- *Contains a high volume of flammable compounds*. Flammable compounds include xylene and toluene, which are found in certain paints, and polyurethane, which is found in varnish. (Water-based paints, such as acrylic or vinyl paints, are non-flammable substances.)
- *Corrosivity*. Highly acidic or highly alkaline aqueous wastes have the ability to corrode other materials and contaminate water sources. For example, the acid in lead–acid batteries has a very low pH (similar to that of sulfuric acid) and poses a significant threat to the environment because it can contaminate water sources.
- *Reactive*. Wastes that are hazardous due to their reactivity may be unstable under normal conditions, may react with water, may give off toxic gases and may be capable of detonation or explosion under normal conditions or when heated. For example, aerosol cans can explode if they are pierced, damaged or overheated, and plasterboard (and similar products such as drywall), which contains gypsum, can produce highly toxic gas if grouped with biodegradable wastes.
- *Toxic*. Toxic substances are hazardous due to the harm they can cause if ingested or absorbed. Antifreeze and brake fluids contain high amounts of diethylene glycol (DEG), which is highly toxic if ingested.
- *Contains potentially carcinogenic substances*. For example, toner and inkjet cartridges for printers contain carcinogenic substances that can disrupt hormonal activity and cause a variety of illnesses.

Other examples of hazardous waste include

- asbestos (contains various toxic chemicals, including carcinogens, which are released if the asbestos is disturbed)
- pesticides
- solvents

- laboratory waste
- cleaning products
- medical waste
- waste electrical and electronic equipment (e.g. fridges, freezers, microwaves and toasters).

5.8.5 Disposing of hazardous waste

Local addresses for hazardous waste disposal services in England and Wales can be found on the UK government website (HMG, 2023c).

Landfill sites should not accept waste, whether inert, non-hazardous or hazardous waste, that does not meet the minimum requirements for waste sent to landfill (see Section 5.11). If waste, including soils, has been classified as hazardous, then waste acceptance criteria testing must be carried out.

The co-disposal of hazardous waste with non-hazardous waste at the same landfill site is prohibited. If the waste is classified as non-hazardous, it only needs waste acceptance criteria testing if it is intended for it to go to an inert landfill site.

However, always check with the relevant landfill site whether there are any additional criteria that must be met under its landfill permit (SOCOTEC, 2023).

5.8.5.1 Waste acceptance criteria (WAC)

The waste acceptance criteria (WAC) must be followed. Details of the WAC are given on the government website (Environment Agency, 2021d).

5.8.5.1.1 Waste not satisfying the WAC. If any hazardous waste is to go to landfill but it does not satisfy the WAC testing, it is called ‘problematic waste’. To dispose of problematic waste to landfill a problematic waste stream form must be completed. The form is obtained from the Environment Agency and returned to its office in the area where the waste is produced (Environment Agency, 2021d).

5.8.5.1.2 Waste acceptance procedures. In addition to characterising waste and complying with the WAC, the Environment Agency requires that a waste acceptance procedure must be in place that shows

- the evidence that is required from producers to confirm that the waste matches its description
- the measures that will be taken to make sure the waste is not contaminated
- the criteria that will be used to decide whether or not to accept the waste (e.g. the results of waste testing)
- a clear explanation of any other criteria used to ensure that only waste that is authorised by your permit is accepted (Environment Agency, 2022c).

5.8.6 Related health and safety law guidance

A breach of environmental law often also involves a breach of health and safety law, particularly in relation to a breach of the Health and Safety at Work Act 1974 (as amended).

There are various health and safety regulations to ensure that hazardous materials are dealt with safely. However, they are not part of environmental law. Examples of such regulations are the Control of Asbestos Regulations 2012 (see Section 7.4, Chapter 7) and the Control of Substances Hazardous to Health Regulations 2002 (see Section 7.3, Chapter 7). Information and guidance about Control of Asbestos Regulations 2012 and the associated approved code of practice and how to manage hazardous materials is given on the Health and Safety Executive website (HSE, 2013, 2023a, 2023b).

Under the Control of Substances Hazardous to Health Regulations

- All personnel should be trained to be aware of and apply proper waste disposal.
- Waste should be classified and separated.
- Appropriate bins and containers should be used to store waste.
- A licensed carrier should be engaged to provide a waste collection schedule.
- A consignment/waste transfer note must be completed for each disposal.
- All documents should be kept on file for at least three years.

5.8.6.1 Safety and storage

- Waste must be clearly labelled 'waste'. An inventory of any waste that is stored on the site or property under your control must be kept and maintained.
- There should be a designated on-site area to store all waste disposal bins.
- Hazardous and non-hazardous waste should be kept separate from each other.
- All bags or bins must be tightly sealed.
- There must be regular waste collections to ensure that bins are not overflowing with waste.
- Regular maintenance checks of the storage area must be carried out to ensure that no bins/containers have become damaged.

Yellow hazardous waste bags should be used for hazardous waste such as dressings and wipes, bandages and personal protective equipment.

Cytotoxic/cytostatic waste bins are available in various sized (2.5–50 litres) products for the disposal of items such as blister packs, medicinal vials and patches.

Hazardous waste wheelie bins are available in a range of sizes (120–1100 litres) and colours. Different colours can be used to for different types of waste.

Intermediate bulk containers (IBCs) are used to store up to 1000 litres of hazardous liquid waste, including chemical waste and sludge/slurries. For guidance on the storage of chemical waste see *Chemical Warehousing: The Storage of Packaged Dangerous Substances* (HSE, 2009).

5.8.7 The Environment Act 2021

Section 60 of the Environment Act 2021 makes changes to section 62 of the Environmental Protection Act 1990 regarding the management of hazardous waste in England and Wales. Section 60(1) states that the relevant national authority (in England, the Secretary of State; and in Wales, the Welsh ministers) may, by regulations, make provision for, about or connected with the regulation of hazardous waste in England and Wales.

This means that it is up to those authorities to determine how and when to exercise the power given to them by this 2021 Act to make changes to existing law. There is no certainty about what the changes will be and how quickly they will come into effect. However, a two-year transition period is expected for the most significant changes regarding waste to start coming into law, which is expected in 2023 and 2024.

Section 62ZA of the Environmental Protection Act 1990 states the following regarding the changes that may be made by the national authorities.

> (2) Provision that may be made in the regulations includes provision
> - (*a*) prohibiting or restricting any activity in relation to hazardous waste
> - (*b*) for the giving of directions by waste regulation authorities with respect to matters connected with any activity in relation to hazardous waste
> - (*c*) imposing requirements about how hazardous waste may be kept (including requirements about the quantities of hazardous waste which may be kept at any place)
> - (*d*) about hazardous waste that originated outside England or Wales
> - (*e*) about the registration of hazardous waste controllers or places where activities in relation to hazardous waste are carried out
> - (*f*) for the keeping of records by hazardous waste controllers
> - (*g*) for the inspection of those records by waste regulation authorities or specified persons
> - (*h*) for the provision by hazardous waste controllers of copies of, or information derived from, those records to waste regulation authorities or specified persons
> - (*i*) for hazardous waste controllers to inform waste regulation authorities, or specified persons, when carrying out activities in relation to hazardous waste
> - (*j*) about the circumstances in which waste which is not hazardous waste, but which shares characteristics with hazardous waste, is to be treated as hazardous waste
> - (*k*) for, about or connected with criminal offences
> - (*l*) for, about or connected with the imposition of civil sanctions.
>
> (3) The regulations may not provide for an offence to be punishable
> - (*a*) on summary conviction, by imprisonment, or
> - (*b*) on conviction on indictment, by a term of imprisonment exceeding two years.
>
> (4) For the purposes of this section 'civil sanction' means a sanction of a kind for which provision may be made under part 3 of the Regulatory Enforcement and Sanctions Act 2008 (fixed monetary penalties, discretionary requirements, stop notices and enforcement undertakings).
>
> (5) The regulations may make provision for, about or connected with the imposition of a sanction of that kind whether or not

(*a*) the conduct in respect of which the sanction is imposed constitutes an offence, or
(*b*) the person imposing it is a regulator for the purposes of part 3 of the Regulatory Enforcement and Sanctions Act 2008.

5.8.7.1 Supervision and records

Section 62ZA of the Environmental Protection Act 1990 states that

(6) The regulations may also include provision
(*a*) for the supervision by waste regulation authorities
(i) of activities in relation to hazardous waste, or
(ii) of hazardous waste controllers
(*b*) about the keeping of records (which may include registers of hazardous waste controllers and places where hazardous waste may be kept or processed) by waste regulation authorities
(*c*) as to the recovery of expenses or other charges for the treatment, keeping or disposal or the re-delivery of hazardous waste by waste regulation authorities or hazardous waste controllers
(*d*) as to appeals to the relevant national authority from decisions of waste regulation authorities.
(7) This section [62ZA] is subject to section 114 of the Environment Act 1995 (which concerns the delegation or reference of appeals to others etc).

5.8.7.2 Meaning of mixing of hazardous waste

Section 62ZA of the Environmental Protection Act 1990 states that

(10) For the purposes of this section, 'mixing' in relation to hazardous waste means

(*a*) diluting it (with any substance)
(*b*) mixing it with other hazardous waste of a different type, or that has different characteristics
(*c*) mixing it with any other substance or material (whether waste or not).

(11) In this section

'activity', in relation to hazardous waste, includes

(*a*) keeping, collecting, receiving, importing, exporting, transporting or producing hazardous waste
(*b*) sorting, treating, recovering, mixing or otherwise processing hazardous waste
(*c*) disposing of hazardous waste in any manner (including providing hazardous waste to another person for the purposes of that person carrying out an activity in relation to it)

(*d*) examining, testing or classifying hazardous waste (including doing any of those things to waste in connection with establishing whether it is hazardous)
(*e*) acting as a broker of, or dealer in, hazardous waste
(*f*) directing or supervising another person in relation to an activity in relation to hazardous waste

'hazardous waste controller' means a person who carries out any activity in relation to hazardous waste

'specified' means specified in the regulations.

5.8.7.3 New definition of hazardous waste and waste list

In section 60(3) of the Environment Act 2021, section 75 of the Environmental Protection Act 1990 has been amended to insert a new definition of 'hazardous waste' (subsections 8A and 8B) and 'waste list' (subsection 8C).

(8A) In the application of this part to England, 'hazardous waste' means
- (*a*) any waste identified as hazardous waste in
  - (i) the waste list as it applies in relation to England, or
  - (ii) regulations made by the Secretary of State under regulation 3 of the Waste and Environmental Permitting etc. (Legislative Functions and Amendment etc.) (EU Exit) Regulations 2020 (S.I. 2020/1540), and
- (*b*) any other waste that is treated as hazardous waste for the purposes of
  - (i) regulations made by the Secretary of State under section 62ZA, or
  - (ii) the Hazardous Waste (England and Wales) Regulations 2005 (S.I. 2005/894).

(8B) In the application of this Part to Wales, 'hazardous waste' means
- (*a*) any waste identified as hazardous waste in
  - (i) the waste list as it applies in relation to Wales, or
  - (ii) regulations made by the Welsh ministers under regulation 3 of the Waste and Environmental Permitting etc. (Legislative Functions and Amendment etc.) (EU Exit) Regulations 2020 (S.I. 2020/1540), and
- (*b*) any other waste that is treated as hazardous waste for the purposes of
  - (i) regulations made by the Welsh ministers under section 62ZA, or
  - (ii) the Hazardous Waste (Wales) Regulations 2005 (S.I. 2005/1806 (W.138)).

(8C) In subsections (8A) and (8B) (above) 'the waste list' means the list of waste contained in the Annex to Commission Decision of 3 May 2000 replacing Decision 94/3/EC establishing a list of wastes pursuant to Article 1(a) of Council Directive 75/442/EEC on waste and Council Decision 94/904/EC establishing a list of hazardous waste pursuant to Article 1(4) of Council Directive 91/689/EEC on hazardous waste (2000/532/EC).

5.8.7.4 Charging schemes – Wales

Section 60(5) of the Environment Act 2021 makes an insertion into section 41(1) of the Environment Act 1995 (power to make charging schemes), before paragraph (d)

(5) In section 41(1) of the Environment Act 1995 (power to make charging schemes), before paragraph (d) insert

'(cc) as a means of recovering costs incurred by it in performing functions conferred by regulations made under section 62ZA of the Environmental Protection Act 1990 (special provision with respect to hazardous waste), the Agency or the Natural Resources Body for Wales may require the payment to it of such charges as may from time to time be prescribed'.

5.9. Transporting and handling waste

The requirements for transporting or handling waste under a permit in the UK were introduced to implement Article 26 of the Waste Framework Directive (2008/98/EC), which was amended in 2018, and modified from 31 December 2020 by section 3D of the retained EU law Waste (England and Wales) Regulations 2011.

There is now a customs border raised between the UK and EU and the Basel Convention regime restricts the export of plastic waste, which is likely to result in greater scrutiny being applied to exports. It also means that for each consignment of waste a customs declaration and related requirements will need to be complied with. The requirements of individual member states will apply.

The Transfrontier Shipment of Waste Regulations 2007 have been amended by the International Waste Shipments (Amendment) (EU Exit) Regulations 2021. The latter provide that Greater Britain will treat Northern Ireland as an EU member state in the context of the transport of waste from Great Britain to Northern Ireland.

5.9.1 After Brexit

Following Brexit, since 31 December 2020, the Basel Convention applies to shipments of waste. The Convention does not permit shipments to countries where environmentally sound management is not guaranteed. Under the Convention, the following notification procedure for the transboundary movements of waste must be applied

- notification
- consent and issuance of a movement document
- transboundary movement
- confirmation of disposal.

The Organisation for Economic Co-operation and Development (OECD) decision C(2001) 107 provides the framework for transboundary movements of waste for recovery between OECD countries (OECD, 2001).

5.9.2 What is transporting waste?

Under the Control of Pollution (Amendment) Act 1989 transport in relation to any controlled waste, includes

- by road or rail
- by air, sea or inland waterway.

It does not include moving waste from one place to another by means of any pipe or other apparatus that joins those two places.

5.9.3 Transporting waste – the duty of care

Anyone concerned with waste must take all reasonable steps to ensure that the waste is

- managed properly, preventing unauthorised or harmful deposit, treatment or disposal
- recovered or disposed of safely
- does not escape and does not cause harm to human health or pollution of the environment
- only transferred to someone who is authorised to receive it
- accurately described when it is transferred to another person.

The duty applies to any person who produces, imports, keeps or manages controlled waste or who as a broker has control of such waste.

An authorised transport purpose is the

- transport of controlled waste between different places within the same premises
- transport to a place in Great Britain of controlled waste which has been brought from a country or territory outside Great Britain and which has not landed in Great Britain until it arrives at that place
- transport by air or sea of controlled waste from a place in Great Britain to a place outside Great Britain.

5.9.4 Carriers of controlled waste – system and reform

Under the current system, the following must be registered.

- Carriers of controlled waste (section 1, Control of Pollution (Amendment) Act 1989). Although the term 'carrier' is not defined in legislation, it refers to someone who transports government-controlled waste.
  The term 'controlled waste' refers to domestic, commercial or industrial waste (part II, Environmental Protection Act 1990).
- Brokers of, or dealers in, controlled waste (regulation 25, Waste (England and Wales) Regulations 2011). They are defined in Article 3 of the Waste Framework Directive (2008/98/EC) as those involved in 'any undertaking arranging the recovery or disposal of waste on behalf of others, including those who do not take physical possession of the waste'. A 'waste dealer' means a legal or natural person who, in their own name and on their own account, purchases and sells waste, including those who do not take physical possession of the waste.

The Waste (England and Wales) Regulations 2011 introduced a two-tier registration system for waste carriers, brokers and dealers

- Upper tier – professional waste carriers, brokers and dealers. These must register every three years.
- Lower tier – 'specified persons'. This tier includes a carrier carrying waste from its own business. Its single registration lasts indefinitely. No fee is payable for lower-tier

registration, and the lower tier benefits from less strict enforcement (see regulation 31, Waste (England and Wales) Regulations 2011).

5.9.4.1 Registration and exemption from registration

The following are not required to register as a carrier of controlled waste under the Control of Pollution (Amendment) Act 1989 (see regulation 26, Waste (England and Wales) Regulations 2011).

- A carrier who is a specified person, such as a charity or a waste collection authority (see part II, Environmental Protection Act 1990), and who does not normally and regularly transport controlled waste.
- An operator of marine shipping (i.e. a vessel, aircraft, hovercraft, floating container or vehicle loaded with waste) that requires a marine licence under the Food and Environment Protection Act 1985 or the Marine and Coastal Access Act 2009 (or that is exempt from that requirement).
- A carrier which only transports waste produced by the carrier itself, except where it is construction or demolition waste (and 'construction' includes improvement, repair and alteration).
- A carrier which only transports, a broker which only arranges for the recovery or disposal of, or a dealer which only deals in

  - animal by-products
  - waste from a mine or quarry
  - waste from premises used for agriculture.

5.9.5 Registration procedure

Registration as a waste carrier (or broker or dealer) or as an exempted carrier is required under both the Waste (England and Wales) Regulations 2011 and the Control of Pollution (Amendment) Act 1989. A carrier, broker or dealer can apply to register in writing or online. The regulator has a duty, where possible, to reduce the administrative burden by using the records it already holds to obtain information for the registration process (regulation 46, Waste (England and Wales) Regulations 2011).

However, the regulator can refuse to register an application if it undesirable for the applicant to be authorised to transport controlled waste or to act as a broker or dealer. This is the case where the applicant or other relevant person has committed an offence under the various environmental legislation listed in the Waste (England and Wales) Regulations 2011 (e.g. where a carrier, broker or dealer has failed to register under the Control of Pollution (Amendment) Act 1989 or regulation 25 of the 2011 Waste Regulations (as applicable), or has committed an offence under section 33 of the Environmental Protection Act 1990). In relation to individuals, the Rehabilitation of Offenders Act 1974 results in convictions being spent after five years. There is no corresponding provision where an offence is committed by a corporate body. The Waste (England and Wales) (Amendment) Regulations 2014 broaden the range of relevant convictions.

Carriers, brokers or dealers have 28 days to tell the regulator of any change in circumstances that will affect their registration (regulation 30 of the 2011 Waste Regulations).

5.9.5.1 Refusal of registration

Where a carrier is refused registration or its registration is revoked (regulation 32 of the Waste (England and Wales) Regulations 2011), it can appeal. An appeal to the Secretary of State (section 4 of the Control of Pollution (Amendment) Act 1989) must be lodged within 28 days of the refusal or revocation (regulation 33, Waste (England and Wales) Regulations 2011).

The regulator will undertake periodic inspections of carriers, brokers and dealers (regulation 34 of the 2011 Waste Regulations).

Part 10A of the Waste (England and Wales) Regulations 2011 allows a person to produce their authority to transport waste up to five working days after the request, rather than seven calendar days.

5.9.5.2 Waste information notes

The Environment Agency requires all waste transfer notes and hazardous waste consignment notes to include a declaration that the carrier, broker, or dealer has taken all reasonable measures to apply the waste hierarchy when waste is transferred.

The Waste (England and Wales) (Amendment) Regulations 2014 came into force on 6 April 2014. It states that invoices and receipts can now be provided rather than waste transfer notes as long as the same information that was contained in the transfer notes is recorded on the invoices, and that waste carriers are allowed up to five days to produce their waste registration certificate, meaning they do not need to carry their registration document at all times while operational.

5.9.5.3 Duty of care consignment declaration

A duty of care for waste applies to waste carriers, brokers and dealers (as it does to anyone else handling controlled waste) to take all reasonable steps

- to ensure that the waste is not disposed of unlawfully, without a permit or in breach of any permit
- to ensure that the waste is not treated, kept or disposed of in a way that causes pollution or harm
- to prevent the escape of the waste from their control or that of any other person
- to prevent any other person committing an offence (i.e. disposing of controlled waste, or treating or storing it) under section 33 of the Environmental Protection Act 1990.

The duty of care requires that waste holders should ensure that their waste is managed at an appropriately permitted facility, and ensure that those managing their waste do so in accordance with permit conditions. It places an obligation on all ‘holders’ of waste to meet certain requirements that help ensure protection of the environment from the point of waste production to its final recovery or disposal (section 34 of the Environmental Protection Act 1990, effective from April 1992).

The waste may only be transferred to an authorised person for authorised transport purposes. It must be accompanied by a written description that enables the transferee to know enough about the waste to deal with it properly and avoid breaching their permit or section 33(1) of the EPA 1990 or of the duty of care owed under Section 34, EPA 1990.

The case described in Box 5.2 demonstrates that the statutory duty of care requires that when a waste holder transfers waste to someone else in the waste chain they have taken all reasonable steps to ensure the waste is managed correctly until its final disposal or other end (e.g. recovery).

Box 5.2 *Walker and Son (Hauliers) Ltd v Environment Agency* [2014] EWCA Crim 100

Land purchased by Walker and Son (Hauliers) Ltd (Walker) for redevelopment was used by Bloom (Plant) Limited, which had been contracted to demolish empty buildings on the site, as a waste transfer station. The local council concluded that Bloom was conducting an illegal waste operation without an environmental permit. Bloom pleaded guilty to various waste offences. Walker was charged with knowingly permitting the operation of a waste facility without a permit. The company pleaded guilty. It then appealed, arguing that it had not been aware of the need for, or omission of, the permit.

The Court of Appeal had to decide whether the defendant was guilty of 'knowingly permitting' the operation of a regulated facility on their land without an environmental permit, contrary to regulation 38(1)(a) of the Environmental Permitting (England and Wales) Regulations 2007. It was argued by Bloom that the prosecution had to show that it knew that the operations were not authorised by an environmental permit.

The court rejected this argument and dismissed the appeal. It held that the words 'knowingly permit' related to knowledge of the facts and not to the existence of the environmental permit. It also held that the prosecution did not have to show that a defendant knew that the matters of which it was aware were not permitted. The prosecution only had to establish that they knew waste operations were taking place, that they had allowed, or failed, to prevent them, and that the operations were not being performed in accordance with an environmental permit. The words 'knowingly permit' did not relate to the existence or scope of conditions attached to the permit, and there was no defence based on the exercise of due diligence.

5.9.5.4 Reform – consultations

A consultation on the reform of the waste carrier, broker and dealer registration system in England was opened in January 2022 by Defra (2022e). The consultation document proposes to change the terminology 'carrier, broker, dealer' (CBD) to that of 'controller' and 'transporter', even though the former set of terms are well established in legislation and in the consultation on the digital waste tracking system. The terminology is also used in relation to waste carried nationally and internationally. Transporters would transport waste, but they would not decide how it is classified or where it goes. Waste controllers would be responsible for classifying waste, as well as deciding where it goes and arranging for a transporter to physically carry it. It would be possible for a permit to be issued for the take up of both roles at once.

The consultation document also proposed moving the current CBD registration system to being regulated under the environmental permitting regime in order to increase its regulation. This means that only legal entities could carry, broker or deal with waste, although there may be a few exemptions.

The consultation closed on 15 April 2022. The transfer to governance under the environmental permitting regime during 2023 into 2024 involves the Pollution Prevention and Control Act 1999 and Environmental Permitting (England and Wales) Regulations 2016 as amended. It means a move from a registration system to a permit system. Organisations which are currently registered will apply for the relevant permit under the new system when their registration is due for renewal. Other organisations and those in the lower tier are required to either register for an exemption or apply for an exemption within 12 months of the start of the new system. To obtain a permit, all are required to satisfy technical competence elements (Defra, 2022b).

5.9.6 Working with the Driver and Vehicle Standards Agency

The Environment Agency and the Driver and Vehicle Standards Agency (DVSA) have signed a memorandum of understanding to stop illegal waste carriers by enabling enforcement against non-compliant waste industry vehicles and carriers. They have formally agreed to share intelligence, to carry out operations to stop illegal waste carriers, and to improve road safety in England.

5.10. Notices under the Waste (England and Wales) Regulations 2011

Compliance notices. The regulator can serve a compliance notice requiring a person to take specific steps within a specific time period to prevent the recurrence or continuation of a breach of

- the waste hierarchy under regulation 12 of the Waste (England and Wales) Regulations 2011
- the requirement to keep different recyclates separate under regulations 13 and 14 of the Waste (England and Wales) Regulations 2011
- the requirement to register as a waste broker or dealer under regulation 25 of the Waste (England and Wales) Regulations 2011, or as a waste carrier under section 1 of the Control of Pollution (Amendment) Act 1989 (regulation 38, Waste (England and Wales) Regulations 2011).

Stop notices. If the regulator considers that the person is acting in breach of any of the categories given above, it can serve a stop notice prohibiting a person from carrying on an activity specified in the notice until that person has taken the steps specified in the notice.

Restoration notices. The regulator can serve a restoration notice requiring a person to take specific steps within a specific time period to restore the position to what it would have been if there had been no breach of regulation 14 of the Waste (England and Wales) Regulations 2011, which concerns anyone who collects, transports or receives waste paper, metal, plastic or glass.

5.10.1 Penalties

The penalty for failing to register, or failing to comply with a compliance notice or a stop notice in relation to registration upon summary conviction is an unlimited fine (section 1 of the Control of Pollution (Amendment) Act 1989 and regulation 42(2) of the Waste (England and Wales) Regulations 2011).

There is also an unlimited fine for failure to comply with a compliance notice, stop notice or restoration notice for breach of regulations 12, 13 or 14 (regulation 42(3) of the Waste (England and Wales) Regulations 2011).

5.10.2 Offences – carriers, brokers and dealers

The enforcement regime for carriers, brokers and dealers registered in the lower tier is less strict than for those registered in the higher tier.

For carriers, brokers and dealers who should be registered in the higher tier, it is an offence to fail to register as a

- carrier of controlled waste (section 1(1) of the Control of Pollution (Amendment) Act 1989)
- broker or dealer of controlled waste (regulations 25 and 42(1) of the Waste (England and Wales) Regulations 2011).

Carriers, brokers and dealers who are specified persons and who should be registered in the lower tier commit an offence if they fail to

- register as a carrier of controlled waste under section 1 of the Control of Pollution (Amendment) Act 1989 – the regulation authority can only issue proceedings if it has already served a compliance notice or stop notice and the offender has failed to comply with the notice (regulation 45, Waste (England and Wales) Regulations 2011)
- comply with a compliance notice or stop notice for failing to register as a broker or dealer (regulation 42(1) Waste (England and Wales) Regulations 2011).

It is an offence for anyone to fail to comply with a compliance notice, stop notice or restoration notice.

Compliance notices and stop notices can be served for failure to register or for breach of the waste hierarchy. Restoration notices can be served.

5.10.3 The environmental permitting regime

Following the Waste and Environmental Permitting etc. (Legislative Functions and Amendment etc.) (EU Exit) Regulations 2020 amendments to Environmental Permitting (England and Wales) (Amendment) (EU Exit) Regulations 2019, waste permits and licences are now managed by environmental permitting (Environment Agency, 2022d). For details on whether a waste permit is needed and how to obtain one, see Section 2.12 in Chapter 2.

5.11. Coding of waste and disposal to landfill

It is a legal requirement that waste should be classified and assessed before it is collected, disposed of or recovered, and therefore, before it goes to landfill. Waste codes for various types of waste must be used for construction and demolition waste, and the codes must be stated on all waste documentation and records. The codes are given on government websites together with guidance on how they should be used (HMG, 2023b; Natural Resources Wales *et al.*, 2021).

Therefore, before waste can go to landfill it must be classified, and this must also include whether the waste is hazardous or non-hazardous.

Each component of mixed waste to be disposed of to landfill must be classified, and it must be described separately and a detailed characterisation of it provided.

Waste information (formerly, the transfer note) must be completed together with a consignment note before it can go to landfill. However not all types of waste can go to landfill.

5.11.1 Waste that cannot be sent to landfill

The government website states that waste which cannot be sent to landfill includes (Environment Agency, 2021e)

- any liquid waste, including waste water but excluding sludge
- waste that would be explosive, corrosive, oxidising, flammable or highly flammable
- infectious medical or veterinary waste
- chemical substances from research and development whose effects are not known
- whole or shredded used tyres, except for bicycle tyres and tyres with a diameter of more than 1400 mm
- waste paper, metal, plastic or glass that has been separately collected to prepare it for reuse or recycling
- residual waste from separately treated collected waste paper, metal, plastic or glass unless that landfill site is permitted to accept it
- gypsum-based waste (e.g. plasterboard) cannot be sent to a landfill cell that accepts biodegradable waste.

A list of landfill sites currently authorised by the Environment Agency under the environmental permitting regulations is given on the government website (Environment Agency, 2023).

Hazardous waste is regarded as problematic waste if it cannot meet the government's waste acceptance criteria. It will require the completion of a problematic waste stream request form, which should be sent to the local Environment Agency office in the area where the waste was produced (Environment Agency, 2021e).

5.11.2 Waste treatment before landfill

Waste must be treated before it goes to landfill. In addition, a basic characterisation or level 1 waste assessment must be carried out so that a decision can be made regarding which class of landfill site the waste must be sent to. The waste must meet the waste acceptance criteria (WAC) and waste acceptance procedures. Testing of the waste may be required.

More details about the above waste management requirements are given on the government website (Environment Agency, 2021e), and the current WAC requirements can be found by contacting the local Environment Agency office.

5.11.3 Landfill tax

The landfill tax is an environmental tax under the Environmental Permitting (England and Wales) Regulations 2016 (as amended) on any and all waste deposited at a landfill site, whether or not the site is licensed to take waste. All landfill sites must be licensed by the government. The tax was created to deter businesses and individuals from sending waste to landfill sites, which does not promote a circular economy. A landfill site is used to compact and store non-recyclable solid waste, and so is not a sustainable waste disposal method. The aim of the tax is, therefore, to encourage recycling or the use of eco-friendly materials.

5.11.3.1 Who pays the landfill tax?

The controller of a licensed landfill site, the licenser or the permit holder, or whoever should have a permit for sites where material is disposed of, has to pay the tax.

A payer of the landfill tax is required by law to register with HM Revenue and Customs (HMRC). A person who makes, knowingly causes or knowingly facilitates a disposal of waste at an unauthorised site is also liable to pay landfill tax.

5.11.3.2 What does the landfill tax apply to?

The landfill tax applies to all waste disposed of by way of landfill at a licensed landfill site on or after 1 October 1996, unless the waste is specifically exempt. For example, the following waste may be exempt

- dredgings – material removed from water
- mining and quarrying material
- pet cemeteries
- filling of quarries
- waste from visiting forces.

The Landfill Tax (Qualifying Material) Order 2011 lists qualifying material subject to the lower rate of landfill tax. Under the Landfill Tax (Miscellaneous Provisions) Regulations 2018, from April 2018 the landfill tax is applied to disposals of material at sites operating without the appropriate disposal permit. The tax also applies to disposals made prior to 1 April 2018, which are still on the site on 1 April 2018. The objective of this policy was to deter non-compliance, and to reinforce the principle of 'the polluter pays'. Detailed guidance on whether the tax applies to your waste, is given in 'Excise Notice LFT1: a general guide to landfill tax' (HMRC, 2022a).

5.11.3.3 Landfill tax rates

The landfill tax is charged by weight and the rates increase on 1 April every year. There are two rates (the prices per tonne are valid from 1 April 2022) (HMRC, 2022b; OBR, 2023)

- lower rate (£3.15 per tonne) – inert or inactive waste (e.g. rocks or soil)
- standard rate (£98.60 per tonne) – all other waste.

Landfill tax that has already been paid may be reclaimed when

- a customer is unable to pay you (bad debt)
- material you accept is later removed to another landfill site
- landfilled material is later removed for recycling, incineration or reuse.

Box 5.3 *Her Majesty's Customs and Excise v Devon Waste Management Ltd & Ors* [2021] EWCA Civ 584

'Black bag' waste material ('fluff') disposed of at landfill sites is used by operators to protect the geomembrane liner of landfill cells and prevent leachate from decomposing waste seeping into the surrounding earth. This case concerned whether materials placed in a landfill cell to protect the membrane lining the cell were subject to landfill tax. The legal issue was whether the landfill site operators disposed of the fluff or the EVP as waste because they disposed of it with the intention of discarding it.

The Court of Appeal allowed HMRC's appeal from the Upper Tribunal's decision that the site operators were 'using' the fluff and the EVP, and that this meant that they did not have any intention to discard it. The Court agreed with the decision of the First Tier Tribunal (FTT) that the disposal of fluff and EVP was made with the intention of discarding the material as waste, despite its use as a protective layer in the landfill cell. The FTT had therefore applied the statutory provisions correctly.

The use that the taxpayers (Devon Waste Management Ltd) had made of the fluff and EVP was insufficient to negate their obvious intention to discard the material. According to the legislation, if the material was being 'discarded', it was taxable.

Lady Justice Rose helpfully set out the following non-exhaustive list of factors that should be considered when assessing who is the relevant disposer, and whether their intention at the time was to discard the material

- whether the material is placed somewhere within the perimeter of the landfill site but not being placed in the cell
- whether it is processed in some way by or on behalf of the landfill site operators
- whether it is separated out from the main body of waste and stored for a time or, conversely, whether it is placed in the cell immediately or soon after it arrives at the landfill site
- whether it is put into the cell with the expectation that it will stay there permanently
- whether there has been a passage of title to the disposer
- the economic circumstances surrounding the acquisition of the materials in question – who paid whom for the material, and whether the disposer would need to buy in alternative material if there was not enough of the material in dispute
- the practicality of applying or disapplying the tax to the material in question.

References

Statutes

Control of Pollution (Amendment) Act 1989
Environment Act 1995
Environment Act 2021
Environmental Protection Act 1990
European Union (Withdrawal Act) 2018
Food and Environment Protection Act 1985
Health and Safety at Work etc. Act 1974
Marine and Coastal Access Act 2009
Pollution Prevention and Control Act 1999
Regulatory Enforcement and Sanctions Act 2008
Rehabilitation of Offenders Act 1974

Regulations and orders

Control of Asbestos Regulations 2012
Control of Substances Hazardous to Health Regulations 2002
Environmental Permitting (England and Wales) (Amendment) (EU Exit) Regulations 2019
Environmental Permitting (England and Wales) Regulations 2016
The Environmental Permitting (England and Wales) Regulations 2007
The Environmental Targets (Residual Waste) (England) Regulations 2023
The Hazardous Waste (England and Wales) Regulations 2005
The Hazardous Waste (Wales) Regulations 2005
The International Waste Shipments (Amendment) (EU Exit) Regulations 2021
The Landfill Tax (Miscellaneous Provisions) Regulations 2018
The Landfill Tax (Qualifying Material) Order 2011
The Regulation (EC) No 1013/2006
The Waste and Environmental Permitting etc. (Legislative Functions and Amendment etc.) (EU Exit) Regulations 2020
The Waste (Circular Economy) (Amendment) Regulations 2020
The Waste (England and Wales) Regulations 2011
The Waste (Miscellaneous Amendments) (EU Exit) (No. 2) Regulations 2019
The Waste Regulations (England and Wales) 2011
Transfrontier Shipment of Waste Regulations 2007
Waste (England and Wales) (Amendment) Regulations 2014

Directives

Waste Framework Directive 2008/98/EU

Case law

Her Majesty's Customs and Excise v Devon Waste Management Ltd & Ors [2021] EWCA Civ 584
Mountpace Ltd v London Borough of Haringey [2012] EWHC 698 (Admin)
Walker and Son (Hauliers) Ltd v Environment Agency [2014] EWCA Crim 100

Websites

Adams K (2022) *ZAP: Zero Avoidable Waste in Construction*. https://asbp.org.uk/wp-content/uploads/2022/07/ZAP-Deliverable-1-Report-final-v2-for-upload.pdf (accessed 09/03/2023).

Defra (Department for Environment, Food and Rural Affairs) (2020a) Circular Economy Package policy statement. Defra, Welsh Government, Scottish Government, and Department of Agriculture, Environment and Rural Affairs (Northern Ireland). https://www.gov.uk/government/publications/circular-economy-package-policy-statement (accessed 09/03/2023).

Defra (2020b) *Environmental Permitting: Core Guidance. For the Environmental Permitting (England and Wales) Regulations 2016 (SI 2016 No 1154)*. https://assets.publishing.service.gov.uk/government/uploads/system/uploads/attachment_data/file/935917/environmental-permitting-core-guidance.pdf (accessed 09/03/2023).

Defra (2021a) 25 Year Environment Plan. https://www.gov.uk/government/publications/25-year-environment-plan (accessed 09/03/2023).

Defra (2021b) *Waste Management Plan for England*. https://assets.publishing.service.gov.uk/government/uploads/system/uploads/attachment_data/file/955897/waste-management-plan-for-england-2021.pdf (accessed 09/03/2023).

Defra (2021c) Legal definition of waste guidance. Defra, Welsh Government, Environment Agency and Natural Resources Wales. https://www.gov.uk/government/publications/legal-definition-of-waste-guidance (accessed 09/03/2023).

Defra (2022a) *Resource Efficiency and Waste Reduction Targets. Detailed Evidence Report*. https://consult.defra.gov.uk/natural-environment-policy/consultation-on-environmental-targets/supporting_documents/Resource%20efficiency%20and%20waste%20reduction%20targets%20%20Detailed%20evidence%20report.pdf (accessed 09/03/2023).

Defra (2022b) UK statistics on waste. https://www.gov.uk/government/statistics/uk-waste-data/uk-statistics-on-waste#recovery-rate-from-non-hazardous-construction-and-demolition-cd-waste (accessed 09/03/2023).

Defra (2022c) Mandatory digital waste tracking. https://www.gov.uk/government/publications/digital-waste-tracking-service/mandatory-digital-waste-tracking (accessed 09/03/2023).

Defra (2022d) Introduction of mandatory digital waste tracking. https://consult.defra.gov.uk/environmental-quality/waste-tracking (accessed 09/03/2023).

Defra (2022e) Consultation on the reform of the waste carrier, broker, dealer registration system in England. https://consult.defra.gov.uk/eq-resources-and-waste/consultation-on-cbd-reform (accessed 09/03/2023).

Defra (2023) Corporate report Environmental Improvement Plan 2023. https://www.gov.uk/government/publications/environmental-improvement-plan (accessed 10/05/2023).

Defra and EA (Environment Agency) (2018) Resources and waste strategy for England. https://www.gov.uk/government/publications/resources-and-waste-strategy-for-england (accessed 09/03/2023).

DLUHC (Department for Levelling Up, Housing and Communities) (2014) National planning policy for waste. https://www.gov.uk/government/publications/national-planning-policy-for-waste (accessed 09/03/2023).

Environment Agency (2016) Waste collection authority separate collection arrangements: survey results. https://www.gov.uk/government/publications/waste-collection-authority-separate-collection-arrangements-survey-results (accessed 09/03/2023).

Environment Agency (2021a) Quality protocols and resources frameworks: rules for all end of waste frameworks. https://www.gov.uk/government/publications/quality-protocols-qps-rules-for-all-qps (accessed 09/03/2023).

Environment Agency (2021b) Get an opinion from the definition of waste service. https://www.gov.uk/government/publications/get-an-opinion-from-the-definition-of-waste-service (accessed 19/06/2022)

Environment Agency (2021d) Dispose of waste to landfill – Waste characterisation. https://www.gov.uk/guidance/dispose-of-waste-to-landfill#waste-characterisation (accessed 09/03/2023).

Environment Agency (2021e) Dispose of waste to landfill. https://www.gov.uk/guidance/dispose-of-waste-to-landfill (accessed 09/03/2023).

Environment Agency (2022a) Waste carriers, brokers and dealers. Registration and responsibilities. https://uk-nrc.com/wp-content/uploads/2018/10/LIT_7806_7221d9.pdf (accessed 09/03/2023).

Environment Agency (2022b) Check if your material is waste. https://www.gov.uk/guidance/check-if-your-material-is-waste#when-a-material-meets-the-end-of-waste-test (accessed 09/03/2023).

Environment Agency (2022c) Waste acceptance procedures for deposit for recovery. https://www.gov.uk/government/publications/deposit-for-recovery-operators-environmental-permits/waste-acceptance-procedures-for-deposit-for-recovery#:~:text=Your%20waste%20acceptance%20procedures%20must%20set%20out%20the%3A,the%20results%20of%20waste%20testing (accessed 09/03/2023).

Environment Agency (2022d) Landfill and deposit for recovery: aftercare and permit surrender. https://www.gov.uk/government/publications/landfill-epr-502-and-other-permanent-deposits-of-waste-how-to-surrender-your-environmental-permit/landfill-and-deposit-for-recovery-aftercare-and-permit-surrender (accessed 09/03/2023).

Environment Agency (2023) Permitted waste sites – authorised landfill site boundaries. https://www.data.gov.uk/dataset/ad695596-d71d-4cbb-8e32-99108371c0ee/permitted-waste-sites-authorised-landfill-site-boundaries (accessed 09/03/2023).

EPA (Environmental Protection Agency) (2023) Defining Hazardous Waste: Listed, Characteristic and Mixed Radiological Wastes. https://www.epa.gov/hw/defining-hazardous-waste-listed-characteristic-and-mixed-radiological-wastes#FandK (accessed 09/03/2023).

Eurostat (2023) Management of waste excluding major mineral waste, by waste management operations. https://ec.europa.eu/eurostat/cache/metadata/en/env_wasoper_esms.htm (accessed 09/03/2023).

HMG (His Majesty's Government) (2018) *A Green Future: Our 25 Year Plan to Improve the Environment.* https://assets.publishing.service.gov.uk/government/uploads/system/uploads/attachment_data/file/693158/25-year-environment-plan.pdf (accessed 09/03/2023).

HMG (2023a) Hazardous waste – consignment notes. https://www.gov.uk/dispose-hazardous-waste/consignment-notes (accessed 09/03/2023).

HMG (2023b) Classify different types of waste. https://www.gov.uk/how-to-classify-different-types-of-waste (accessed 09/03/2023).

HMG (2023c) Find a local hazardous waste disposal service. https://www.gov.uk/hazardous-waste-disposal (accessed 09/03/2023).

HMRC (His Majesty's Revenue and Customs) (2022a) Excise Notice LFT1: a general guide to landfill tax. https://www.gov.uk/government/publications/excise-notice-lft1-a-general-guide-to-landfill-tax/excise-notice-lft1-a-general-guide-to-landfill-tax (accessed 09/03/2023).

HMRC (2022b) Landfill tax rates. https://www.gov.uk/government/publications/rates-and-allowances-landfill-tax/landfill-tax-rates-from-1-april-2013 (accessed 09/03/2023).

HSE (Health and Safety Executive) (2009) *Chemical Warehousing: The Storage of Packaged Dangerous Substances.* https://www.hse.gov.uk/pubns/books/hsg71.htm (accessed 09/03/2023).

HSE (2013) *Control of Asbestos Regulations 2012. Approved Code of Practice and Guidance.* https://www.hse.gov.uk/pubns/books/l143.htm (accessed 09/03/2023).

HSE (2023a) Materials storage and waste management. https://www.hse.gov.uk/construction/safetytopics/storage.htm (accessed 09/03/2023).

HSE (2023b) Control of Substances Hazardous to Health (COSHH). https://www.hse.gov.uk/coshh (accessed 09/03/2023).

Natural Resources Wales, Scottish Environment Protection Agency (SEPA), Environment Agency (2021) *Waste Classification. Guidance on the Classification and Assessment of Waste*, 1st edition v1.2.GB. Technical Guidance WM3. https://assets.publishing.service.gov.uk/government/uploads/system/uploads/attachment_data/file/1021051/Waste_classification_technical_guidance_WM3.pdf (accessed 09/03/2023).

OBR (Office for Budget Responsibility) (2023) Landfill tax. https://obr.uk/forecasts-in-depth/tax-by-tax-spend-by-spend/landfill-tax (accessed 09/03/2023).

OECD (Organisation for Economic Co-operation and Development) (2001) C(2001)107. *Revision of Decision C(92)39/FINAL on the Control of Transboundary Movements of Wastes Destined for Recovery Operations*. https://one.oecd.org/document/C(2001)107/En/pdf (accessed 09/03/2023).

SOCOTEC (2023) Waste acceptance criteria. https://www.socotec.co.uk/our-services/environmental-science/waste-acceptance-criteria (accessed 09/03/2023).

Welsh Government (2010) *Towards Zero Waste: Our Waste Strategy*. https://www.gov.wales/towards-zero-waste-our-waste-strategy (accessed 09/03/2023).

Baker F and Charlson J
ISBN 978-0-7277-6645-8
https://doi.org/10.1680/elsc.66458.151

Chapter 6
Water pollution

Francine Baker

6.1. Introduction

It is generally expected that water availability will decrease due to climate change, and that short-duration drought conditions will increase. The way in which water is abstracted needs to be managed to address these challenges. This is another area where the construction industry can assist in the achievement of Sustainable Development Goals and mitigate the effects of climate change.

Water pollution incidents can easily arise from construction activities, especially when there is wet weather. Just stripping the soil of topsoil may cause pollution when it rains, as vegetation loses its run-off protection and the resulting muddy run-off into drains and watercourses can smother aquatic life and plants. The run-off from the use of chemicals such as cement, fuel and solvents can also present a serious risk of water pollution, as can construction on brownfield sites involving risks of pollution from disused storage tanks and disused mines.

6.1.1 Chapter contents summary

This chapter refers to the relevant legislation and policy that must be complied with if construction work is conducted near water, or if water that is being used either intentionally or unintentionally discharges anything other than clean water into the surrounding environment. It concerns current or proposed activities which have an impact on or are affected by groundwater and therefore should be noted by

- developers
- planners
- environmental permit applicants and holders
- water abstractors.

It focuses on the law in England and Wales, and to government policy concerning England. There is some reference to the law in Northern Ireland.

This chapter will refer to developments and activities that may involve water discharge, and groundwater discharge activities, and related offences, penalties, defences. It will discuss the reforms under the Environment Act 2021 regarding water quality and land valuation, the water abstraction regime and licences to abstract, and reforms and the move to the environmental permitting regulations regime.

It will also consider the 2022 reforms to the water abstraction charging scheme. It will consider the role played by partnerships with Natural England, which is available to advise local planning authorities, planners, and developers how they can mitigate the expected increase in nitrogen and phosphorus from a new development and achieve nutrient neutrality.

The above will be considered in the light of the Environment Act 2021, which includes a new requirement for a proposed development to show a biodiversity net gain, as well as the implementation of covenants. These recent developments are discussed in more depth in Sections 3.6.1 to 3.6.4 in Chapter 3 on environment impact assessments.

This chapter should be read in conjunction with Section 2.12 in Chapter 2, which concerns environmental permits for water-related construction activities.

6.2. The plan, legislation and regulators

The government's 25-year plan to improve the environment, published on 11 January 2018 and revised 31 January 2023 (Chapter 5 Section 5.2), sets out how to provide clean and plentiful water, create richer habitats for wildlife, and improve air quality and reduce waste. The plan mainly applies to England. However, it applies to the UK as a whole concerning climate change because the UK government is responsible for the policies and programmes that affect sectors across the UK and internationally (Defra *et al.*, 2019).

The Environment Act 2021 has furthered the plan's progress by requiring that the Secretary of State obtain data about the progress of such plans to monitor whether the natural environment, or particular aspects of it, are improving in accordance with the current environmental improvement plan (section 16(1)). The Secretary of State must also publish any such data before Parliament (sections 16(1) to 16(5)).

One of the ways in which the progress in implementing the plan is being measured is through the Department for Environment, Food and Rural Affairs' (Defra's) indicators framework, which tracks change in the environmental system. Natural England is the government's adviser for the natural environment in England. It works with government body partners and provides advice to planning authorities to support them in making plans and decisions that conserve and enhance the natural environment and contribute to sustainable development (Hughes, 2022).

The following sections in this chapter cover the main legislation in England and Wales in a historical order. You should seek further legal advice to ensure that any action complies with the law.

6.2.1 Brexit exit regulations

The Floods and Water (Amendment etc.) (EU Exit) Regulations 2019 amends the floods and water legislation. Part 2 amends primary legislation, part 3 amends secondary legislation, and part 4 amends and revokes certain EU decisions.

6.2.2 Water pollution legislation in England and Wales

The water industry was privatised in 1989 under the Water Act 1989. The regulatory regime for the privatised water industry is principally set out in the Water Industry Act 1991, and amendments made to that Act, notably in 2003 and 2014.

Water Resources Act 1991. The licensing of water abstraction was first introduced in the 1960s. The licensing regime is principally set out in the Water Resources Act 1991, which consolidated existing legislation. This Act regulates water quality and the prevention of water pollution, and enables regulators to act to protect the environment and the needs of water users. It creates water pollution offences based on the 'polluter pays' principle. It set out the functions of the National Rivers Authority (now the Environment Agency), and introduced water quality classifications and objectives.

The legislative regime for flood risk management by the government and other public authorities is set out in various pieces of legislation. The principal primary legislation which concerns the powers and duties of internal drainage boards is the Land Drainage Act 1991.

Water Industry Act 1991. The Water Industry Act 1991 regulates the water and sewerage industries, and the provision of consents for discharge to sewers. It also defines the powers of the Director General of Water Services (now the Water Services Regulation Authority (Ofwat)).

Water Act 2003. The Water Act 2003 provides a modern legislative framework to facilitate both sustainable water resources management and economic growth. It amended the framework for abstraction licensing, made changes to the corporate structure of economic regulation, and extended the scope for competition in the industry to large users.

The Water Environment (Water Framework Directive) (England and Wales) Regulations 2003 introduced a system of river basin management planning with the general aim of achieving good status of surface and ground waters by 2015.

The EU Water Framework Directive (2000/60/EC) was revoked and replaced in England, Wales and Northern Ireland by

- the Water Environment (Water Framework Directive) (England and Wales) Regulations 2017
- the Water Environment (Water Framework Directive) Regulations (Northern Ireland).

Marine and Coastal Access Act 2009. The Marine and Coastal Access Act 2009 established the Marine Management Organisation. It changed the system for licensing the carrying on of activities in the marine environment, and the way in which marine fisheries are managed at a national and a local level. It also modified the way in which licensing, conservation and fisheries rules are enforced, and provides for the designation of conservation zones.

The Environmental Permitting (England and Wales) Regulations 2016 provide a system for regulating polluting activities through environmental permits. They set out an environmental permitting and compliance regime that applies to a range of activities and industries.

Water Act 2014. The Water Act 2014 introduced reforms to the water industry with the aims of making the industry more innovative and responsive, and increasing resilience to flooding and drought. The Act gave Ofwat new powers to make rules about charges and charge schemes. It included new provisions for flood insurance in areas of high flood risk and drainage boards.

Environment Act 2021. The Environment Act 2021 provides powers to make regulations amending existing water quality legislation and places a duty on the Secretary of State to set long-term targets for water.

6.2.3 The water regulators and advice

The regulators that deal with water pollution incidents and enforcement are the Environment Agency for England, and Natural Resources Wales for Wales. For both countries, the Water Industry Act 1991 requires that local authorities monitor the quality and adequacy of water, and may require a water supplier to take appropriate action. The Drinking Water Inspectorate and the water companies (under the Water Supply (Water Fittings) Regulations 1999) have similar but additional roles, as well as enforcement.

The Water Regulations Approval Scheme (WRAS) is funded by water suppliers to provide advice about water regulations such as the Water Supply (Water Fittings) Regulations 1999. The WRAS website provides a list of water suppliers (WRAS, 2023).

6.3. Developments and activities which may involve water discharge

- New housing developments – wastewater.
- Livestock housing and agricultural development, including slurry pits, silage clamps and so on.
- Engineering works within water (e.g. hydroelectric schemes, bridges, weirs and aquaculture).
- Energy generation – energy from waste, wind farms, single wind turbines, power stations and solar energy plants.
- Anaerobic digestion facilities.
- Waste management activities.
- End-of-life vehicle storage and treatment, vehicle fuel filling stations and car wash facilities.
- Cemeteries.
- Landfill sites.
- Infrastructure developments (e.g. roads and pipelines).

6.3.1 Water discharge and groundwater discharge activities

Any water discharge activity which involves the discharge or entry into inland freshwaters, coastal waters or territorial waters of any poisonous, noxious or polluting matter, waste matter, trade effluent or sewage effluent without a permit is prohibited. Activities which result in water discharge and groundwater discharge activities are only permitted through the environmental permit regulations. This includes wastewater from used business and commercial and industrial sources. It includes poisonous, noxious or polluting matter, polluting substances, waste matter, trade or sewage effluent.

Guidance on discharges to surface water and groundwater and environmental permits is given on the government website (Environment Agency and Defra, 2018, 2022a). Note that all government documents are subject to updates. For more details about environmental permits see Chapter 2 Section 2.12.

A groundwater discharge activity concerns the discharge of a pollutant that results in the direct input, or can lead to the indirect input, of that pollutant into groundwater. The location of the activity may present a risk of environmental damage when conducting the following activities

- deep drilling or boreholes
- abstraction of groundwater/surface water
- mineral extraction dewatering
- construction involving piling.

Consider whether the works are to be conducted in or within 10 metres of any waterway which risk affecting any waterway as a result of sediment release into watercourses, discharge of effluent, or spillages of oils, fuels, cement, chemicals or other pollutants.

6.3.2 Modification to the watercourse channel

Consider whether there will need to be any temporary or permanent modification to the watercourse channel or whether the development will be within 250 metres of a private water supply (spring, well or borehole) used for drinking water or for supply for commercial activity. The Environment Agency's flood map can be used to help identify environmental risks as well as flood risks (Defra and Environment Agency, 2022). The temporary blockage of water-courses is regulated by the Standard Rules SR2015 No. 30. Temporary diversion of a main river, created under Chapter 4, of the Environmental Permitting (England and Wales) Regulations 2016 (Environment Agency, 2015), and should be referred to for guidance.

6.3.3 Environment Agency regulatory position statements

The Environment Agency will not normally take enforcement action for operating without a permit for a waste operation provided there has been compliance with its relevant regulatory position statement (RPS) and your activity does not, and is not likely to, cause environmental pollution or harm human health (Defra, 2021a). However, always check with the Environment Agency for updates. For a discussion of environmental permits see Section 2.12 in Chapter 2.

6.3.4 Treating and using wastewater that contains concrete and silt

An environmental permit to treat or use wastewater that contains concrete and silt at construction sites is not needed if the conditions in the Environment Agency's regulatory position statement (RPS 235) are followed (Environment Agency, 2020).

6.3.5 Temporary discharge of uncontaminated water

The Environment Agency's RPS 261 deals with the temporary dewatering from excavations to surface water (Environment Agency, 2023). It concerns temporary discharges of uncontaminated water from excavations to surface water without a water discharge activity permit.

It covers the discharge of uncontaminated water (such as rainwater) from excavations at building sites and other excavations. RPS 261 does not change the legal requirement to obtain a water discharge activity permit when you discharge uncontaminated water (wholly or mainly rainwater) from site excavations to surface water. It does not apply to discharges to ground or groundwater. An environmental permit is usually needed if you discharge liquid wastewater into surface water (see Chapter 2 Section 2.12). RPSs are regularly reviewed, and you should check with the Environment Agency for updates and to see if you need to apply for a permit.

6.3.6 Water discharge to a sewer

A sewage discharge is where you release sewage to either to the ground (e.g. in your back garden) or to surface water (e.g. a river, stream, estuary, lake, canal or coastal water) (Environment Agency, 2021). Water can be discharged to the public foul sewer, subject to the sewerage undertaker's consent. See the guidance from the Environment Agency and Defra on the government website about water discharges to a sewer (Environment Agency and Defra, 2022b).

To find out if there is a public foul sewer near your property, contact your local water company. The water companies' maps may not show all their sewers near you, so you may also need to ask your neighbours if their properties are connected to the public foul sewer.

6.3.6.1 New connections to the public sewer

If you need to make a new connection to the public sewer you will need to check with your sewerage undertaker (usually the local water company). If you need to discharge anything other than domestic sewage you will also need to check with the local sewerage undertaker.

The Environment Agency will not give you a permit for a private sewage treatment system if it is reasonable for you to connect to the public sewer. The factors it will consider include considering how close you are to a public foul sewer (Environment Agency and Defra, 2022b).

6.3.7 Offences, penalties, defences and company director liability

Offences. It is an offence to cause or knowingly permit a water discharge activity or groundwater activity except under, and to the extent authorised by, an environment permit (see Section 2.12 in Chapter 2). The regulator can exercise the relevant powers and apply the available penalties under regulation 38 of the Environmental Permitting (England and Wales) Regulations 2016.

It is also an offence to fail to comply with the requirements of an enforcement notice or of a prohibition notice, suspension notice, landfill closure notice, mining waste facility closure notice, flood risk activity emergency works notice or a flood risk activity remediation notice.

There may be a warning, formal caution or prosecution.

Penalties. If a person is tried and convicted in a magistrates' court, they could be fined and/or sentenced to up to 12 months in prison. If they are tried and convicted in a Crown Court, there could be an unlimited fine and/or a sentence of up to five years in prison.

Defences. If the activity was done in an emergency to avoid danger to human health, there may be a defence, provided that reasonable steps were taken to minimise pollution and the regulator was promptly notified. It may also be a defence if the discharge was from a mine that was abandoned before 2000.

Company officers' liability. Under regulation 41 of the Environmental Permitting (England and Wales) Regulations 2016, directors, managers and secretaries may be held liable for offences that their company (body corporate) is guilty of, and which are proven to have been committed with their consent or connivance, or are attributable to their neglect. Members of a body corporate may also be liable if they manage the company affairs. Again, there may be a warning, a formal caution or Prosecution.

Civil sanctions are not available for this offence.

6.4. Permit objections and refusals

Anyone who objects to the grant of a permit by the Environment Agency (EA) has no legal right to appeal the decision. It may be possible to apply to the High Court for a Judicial Review; however, this is costly and legal advice should be obtained.

If the EA refuses to grant a permit, there may be an appeal which is usually conducted by planning inspectors. Third parties are allowed to make representations about the appeal. Advice can be obtained from the Planning Inspectorate, and more details are given in Chapter 2.

6.5. Overlap with environmental damage regime

There is some overlap of the environmental permitting regime (Environmental Permitting (England and Wales) Regulations 2016) (see Chapter 1) and the environmental damage regime, as the regulator may also request action under the Environmental Damage (Prevention and Remediation) (England) Regulations 2015 or Environmental Damage (Prevention and Remediation) (Wales) Regulations 2009.

Schedule 2 of the latter regulations hold the operator strictly liable for environmental damage to land, water and protected species and habitats caused by certain polluting activities. This includes sites with environmental permits, water abstraction, storage of chemicals and waste operations.

6.6. Large reservoir constructions

The Secretary of State, in exercise of the powers conferred on them by sections 12A(1) and 12A(2) of the Reservoirs Act 1975, made the Flood Plan (Reservoirs Emergency Planning) Direction 2021 to undertakers in relation to large raised reservoirs in England.

6.6.1 Who is an undertaker?

An undertaker could be the people carrying on the undertaking (i.e. developer), the Environment Agency, the Natural Resources Wales or a water undertaker, or the owner or lessee of the reservoir.

6.6.2 What must an undertaker do?

1. An undertaker must prepare a flood plan for each large raised reservoir in England (including any under construction) for which it is responsible, in accordance with this Direction and the section 12A and 12AA of Reservoirs Act 1975 ('the Act').
2. The undertaker of a large raised reservoir that has been designated as high risk must prepare a flood plan within 12 months of the date of the notice serving this Direction.
3. The undertaker of a large raised reservoir that has not been designated as a high-risk reservoir must prepare a flood plan within 12 months of the date of the notice serving this Direction.
4. The undertaker of a large raised reservoir under construction must prepare a flood plan before the construction engineer issues a preliminary certificate for the reservoir in accordance with section 7 of the Act.
5. The undertaker must prepare the flood plan in accordance with such methods of technical or other analysis as may be specified by the Environment Agency.

Matters to be included in the flood plan

6. A flood plan must include as a minimum all the following information
   (*a*) a clear plan of actions that the undertaker will take in order to prevent an uncontrolled escape of water
   (*b*) a clear plan of actions that the undertaker will take in order to control or mitigate an uncontrolled escape of water
   (*c*) where this information has not already been included in the information required by paragraphs (a) or (b), instructions on how to carry out an emergency drawdown of the reservoir
   (*d*) the areas that may be flooded in the event of an uncontrolled escape of water from the reservoir, including reservoir flood risk maps published by the Environment Agency where available.

For more requirements and information about whether you are considered an undertaker or are constructing a large raised reservoir, contact the Environment Agency in your area. There is also general government guidance which provides an Environment Agency guide to risk assessment for reservoir safety management (Defra, 2021d).

6.7. Environment Act 2021 – water quality

Part 5 of the Environment Act 2021 gives the Secretary of State the power to make regulations about the substances to be considered in assessing the chemical status of surface water or groundwater, and to specify standards for those substances or in relation to the chemical status of water bodies.

6.8. Environment Act 2021 – land valuations

Part 5 of the Environment Act 2021 reforms the process of the calculation of the contributions to internal drainage boards (IDBs) by landowners and local authorities to enhance flood and coastal erosion risk management. The Environment Act 2021 amends the Land Drainage Act 1991 to modernise the valuation of land for the purpose of apportioning the costs of IDBs and calculating drainage rates and special levies between agricultural landowners (via drainage rates) and local authorities (via the special levy).

The amendments allow for the government to pass regulations that will enable the necessary updates to valuation calculations (including data sources) to be readily made. These calculations are used by IDBs to correctly apportion their expenses.

6.9. The water abstraction regime

'Water abstraction' simply means the process of extracting water from any natural source, such as a lake, aquifer, river, stream or spring.

6.9.1 Water abstraction regulation

The Water Resources Act 1991 (as amended) is the main legislation governing water abstraction. The licensing regime is found in part II of the Act. The regulator is the Environment Agency in England and the Natural Resources Wales in Wales.

As part of the delivery of the government's long-term goal of clean and plentiful water under its 25-year environment plan, the Water Abstraction Plan 2017 sets out how the government will reform water abstraction management and how this will protect the environment and improve access to water (Defra, 2021c). The key milestones that remain to be achieved include

- 2023 – abstraction and impounding moves into the Environmental Permitting (England and Wales) Regulations 2016 and all existing abstraction and impounding licences become environmental permits (subject to consultation)
- by 2027 – the Environment Agency will have updated all abstraction licensing strategies.

6.9.2 Licence to abstract

Applications for licences are made under section 37 of the Water Resources Act 1991. The Environment Agency must publish a notice of the application and serve it on water undertakers, the harbour or conservation authority, the navigation authority and the drainage board. The public may make representations about the application to the Environment Agency, and no decision will be taken during this time.

The Environment Agency will consider the application under section 38, and it should take into account all relevant circumstances. It has wide powers to impose conditions on the grant of a licence, or it can reject the application.

Under section 39(1), the Environment Agency cannot grant a licence that derogates from another's protected rights without their permission. Under section 39(2), where the abstraction concerns taking from underground strata the Environment Agency must have regard to the requirements of existing lawful uses of the water.

The Environment Agency must have regard to what is an acceptable minimal flow of water under section 40.

Any person who abstracts more than 20 cubic metres of water a day from surface waters or groundwater must obtain an abstraction licence from the Environment Agency or Natural Resources Wales. There are a few exemptions.

Licences may be granted specifying the amount of water that can be abstracted and what it can be used for. However, a licence may only be granted to a person who has a property right to access the land overlying a relevant aquifer or adjoining relevant surface waters.

A licence granted after 2004 has a time limit, usually between 6 and 18 years. However, older licences are not often time-limited. A full abstraction licence is a licence that has been granted for more than 28 days. It can be traded permanently or temporarily with another person. Further information about how to apply for a water abstraction or impounding licence is available on the government website (Environment Agency, 2022a).

Box 6.1 *Harris & Anor v The Environment Agency* [2022] EWHC 2606

This case illustrates that the Environment Agency needs to comply with the Conservation of Habitats and Species Regulations 2017 when investigating whether to provide water abstraction licences. It also illustrates that European Directives concerning nature conservation laws are still applicable post-Brexit.

The High Court held in September 2022 that the Environment Agency had failed to adequately consider the impact of providing licences for water abstractions in a special area of conservation (SAC) in the Norfolk Broads by limiting its investigation to only three sites in the SAC. The SAC covers approximately 22.5 square miles. The court held that the Environment Agency should have considered the effect on the whole of the SAC, and that the Environment Agency had therefore breached its obligations under Article 6(2) of the European Habitats Directive (92/43/EEC) to avoid the deterioration of protected habitats and disturbance of protected species in SACs. Article 6 of the Directive is enforceable in the UK courts, despite Brexit, because its obligations had been recognised in cases decided prior to the UK leaving the EU, and therefore satisfied the test in section 4(2)(b) of the European Union (Withdrawal) Act 2018.

The Environment Agency also breached the Conservation of Habitats and Species Regulations 2017 (part 1, regulation 9(3)), which requires the Agency to have regard to the above-stated obligations.

The court provided its order (remedy) in October 2022. This requires the Environment Agency to devise a plan to avoid continuing to be in breach of its legal obligations. The court stated

> The defendant shall, by 4 pm on 7 December 2022, provide to the claimants details of the measures it intends to take to comply with its duties under Article 6(2) of the Habitats Directive ('Article 6(2)') in respect of the Broads Special Area of Conservation. The details shall include an indication as to the time by which the defendant intends to have completed those measures. It shall also include, so far as practicable, the scientific and technical basis for the defendant's assessment of the measures that are necessary to comply with Article 6(2).

6.9.3 Offence and penalties

The abstraction of water from any source of supply (or to carry out associated works) without a valid licence or where the terms of an abstraction licence are breached is a criminal offence. Any person found guilty of the offence is liable to an unlimited fine.

If the actions of a person cause loss or damage to a third party, there may also be civil liability. The Environment Agency or Natural Resources Wales may decide to impose a civil penalty.

6.9.4 The role of the Secretary of State and appeals

The regulator or the Secretary of State, rather than the Environment Agency, can call in an application for a licence and determine it.

Where the Environment Agency rejects a licence application, the applicant can appeal to the Secretary of State, who can revoke or vary an abstraction licence where they think this is necessary. In certain cases, it may temporarily reduce or stop abstraction under a licence. If the applicant is unhappy with the result of the appeal, there can be a hearing or inquiry.

6.9.5 Compensation

If the Secretary of State varies or revokes a licence, compensation from the Environment Agency or Natural Resources Wales is available to a licence holder if they have either incurred expenditure in carrying out work made abortive by the revocation or variation, or sustained loss or damage directly attributable to the revocation or variation.

However, where the change has been made for certain environmental protection reasons, no compensation is payable.

6.9.6 Environment Act 2021 – reforms to licences

Part 5 of the Environment Act 2021 sets out provisions to secure long-term, resilient water and wastewater services. It amends provisions in the Water Resources Act 1991 so that, on or after 1 January 2028, permanent licences in England can be varied or revoked without compensation where the change is necessary to protect the environment. These provisions apply in England only.

6.9.7 The environmental permitting regime – water abstraction and licences regulation

The government held a 12-week consultation in September 2021 about its proposed changes to the regulatory framework for abstraction and impounding licensing in England so that it moves into the environmental permitting regulations regime (Defra, 2021d). This move is going ahead but it has been delayed until 2023. It is possible that its implementation may be delayed further.

Principles from the water resources legislation that are fundamental to protecting abstractors will be moved into the environmental permitting regulations. The rights, entitlements or conditions of existing licences will not change. However, new applicants or applicants who vary or transfer a transitional permit and who are issued with a new permit after the move will have different rights and conditions. The objective of the move is to modernise abstraction and its management and to streamline the overall regulatory framework.

One permit may be allowed to cover all the operations on one site. This will reduce the administrative burden on operators who hold more than one legal permission with the Environment Agency and conduct multiple activities at the same site.

6.10. Changes to water abstraction charging scheme

A new scheme was implemented by the Environment Agency in April 2022. Under the old scheme, abstractors were charged relative to the potential effect on the water environment of the abstraction covered by the licence. There were two tiers of charges for the application: and the majority paid the lower tier fee (£135), and others the higher tier fee (£1500). In addition, an annual charge was levied that was based on the volume of water authorised to be abstracted.

6.10.1 New scheme application costs

Under the new scheme, from 1 April 2022, any operator who applies for a new licence or to vary an existing licence will be subject to the new charge. The application charge covers the cost of the activities that the Environment Agency carries out to assess and determine an application for an abstraction licence.

It consists of two parts

- the cost of activities that the Environment Agency carries out to assess and determine the application, which is determined according to the activity type, the volume applied for, and the water availability
- if the extra costs of any additional work the Environment Agency must carry out with respect to the application.

So, the new annual charge is based on the cost of the water resource management activities carried out by the Environment Agency.

Tables of the charges can be found in the document *The Environment Agency (Environmental Permitting and Abstraction Licensing) (England) Charging Scheme 2022* which can be accessed via the government website (Environment Agency, 2022b).

6.11. Nutrient pollution and mitigation

Nutrient pollution in watercourses damages freshwater habitats and estuaries and disrupts natural processes, harming the plants and wildlife and ecosystems that we depend on.

Under the authority of the Environment Act 2021, the government's proposed target is to reduce nutrient pollution in the water environment by reducing the phosphorus loading from treated wastewater by 80% by 2037, and reducing the nitrogen, phosphorus and sediment arising from agriculture by 40% by 2037 (Defra, 2022).

Construction activities may produce many types of water pollution. Ordinary day-to-day building activities that involve chemicals such as paint, solvents, fuel and cement may produce significant water pollution. A common source of water pollution on construction sites results from stripping the topsoil from a site, creating silt-laden waters. When it rains, without vegetation to protect the soil from erosion, the water run-off and surface scouring increases, and this is exacerbated by the use of heavy machinery. Soil particles are released and suspended in the air. The muddy water can block drains and watercourses, reducing the light and oxygen available for aquatic creatures and plants. A proposed development within a watercourse may affect the hydromorphology and the quantity and flow of surface waters. In addition, any development that requires excavation and/or changes in the ground level may disrupt groundwater flow paths (Coulter, 2019).

Petroleum-based substances such as petrol, diesel, kerosene and oils are all hydrocarbons. Historical spillages of hydrocarbons have been absorbed into the land, or have arisen from unmapped pipe networks still full of product or from new spillages from refuelling of plant or burst hoses (Environment Agency, 2019).

Alkaline water has the potential to cause greater harm than silt or oil to the aquatic environment when discharged. Alkaline water on construction sites is mostly the result of the washing of concreting plant and tools. Concrete washwater is strongly alkaline (pH of around 12 to 13), and it would take 10 000 litres of pH neutral water to get 1 litre of washwater with a pH of 12 down to an acceptable pH of 8 (Coulton, 2016).

However, the use of mineral acids like sulphuric or hydrochloric to adjust the pH is dangerous. They are dangerous to handle, as is citric acid, and any spillage can produce polluting acidic water. Citric acid also increases the water's biochemical oxygen demand above safe limits. Therefore, carbon dioxide has been considered the best neutralising agent for concrete water (Elmes and Coleman, 2021; Yoo *et al.*, 2017). However, carbon dioxide is a greenhouse gas, and therefore instrumental in climate change. The government is attempting to greatly reduce the production of carbon dioxide and to capture and safely store that being produced.

Developers and managers should therefore consider the potential risks of the proposed activity to the quality and flow of surface water and groundwater, and to fisheries interests.

6.11.1 The role of Natural England and government advisers

Natural England works with and supports planning authorities and developers to effectively implement nutrient neutrality in order to maintain healthy rivers, lakes and estuaries in the vicinity of a development. Wastewater from new housing developments can have detrimental effects and undermine ongoing efforts to recover aquatic sites.

In addition, Natural England has partnered with Defra and the Department for Levelling Up, Housing and Communities (DLUHC) to create tools and guidance using the latest evidence and bespoke catchment calculators to assess a site's current nutrient status and the likely impact of any new development. This will help competent authorities and developers to identify the level of mitigation required to cancel out the additional nutrient pollution expected from a particular project (Natural England *et al.*, 2022).

The package is targeted at the areas where catchments and the legally protected nature sites they contain are under the most pressure from high levels of nutrients. Natural England has already advised various planning authorities, mostly in the south of England, and this has enabled new development to proceed in a sustainable way known as 'nutrient neutrality', by supporting the authorities in making plans and decisions that conserve and enhance the natural environment and contribute to sustainable development (Natural England, 2022).

6.11.2 Mitigation measures

Typical mitigation measures that have been advised involve providing new spaces for nature and recreation, such as the creation of new wetlands, woodland or grasslands, or the installation of environmentally friendly sustainable drainage systems (SUDS).

Box 6.2 Case study

On 20 November 2017 at Leeds Magistrates' Court, Harron Homes Limited, a Leeds-based building company, was fined £120 000 for illegally polluting a watercourse from a Huddersfield construction site in 2015. The company was ordered to pay £8706.71 in legal costs and a £120 victim surcharge.

An officer from the Environment Agency visited the site on 20 November 2015 and saw polluted water flowing out of the entrance of the construction site. The company was also pumping silt-contaminated water from site excavations, which also entered the watercourse.

Harron Homes attempted to control the silt run-off by setting up settlement tanks. Subsequent inspections revealed that this system was inadequate. Silty water was found to be discharging, resulting in further pollution.

A healthy watercourse is expected to have a concentration of suspended solids lower than 30 milligrams per litre of water. However, some discharge samples showed nearly 35 000 milligrams per litre. The discharges had a significant impact on the water quality in the watercourse up to 3 kilometres downstream.

(Edgar, 2018; Environment Agency, 2017)

6.11.3 Water companies upgrade

> In England, 27 water catchments (encompassing 31 internationally important water bodies and protected sites) are in unfavourable status due to nutrient pollution (Defra *et al.*, 2022a).

On 20 July 2022 the government announced plans that included amendments to the Levelling Up and Regeneration bill (UK Parliament, 2022). A public consultation ended on 2 March 2023 (DLUHC, 2022), and the bill is going through Parliament at the time of writing. The bill concerns reforms to national planning policy. It includes plans to impose a new legal obligation on water companies in England to upgrade their wastewater treatment works in 'nutrient neutrality' areas by 2030 using the highest levels of available technology. These upgrades must be made in a way that effectively addresses water pollution at protected wildlife sites and protected areas. The government intends to work with water companies to identify how to accelerate upgrades in the most effective manner.

6.11.4 Plan to reduce water pollution

Natural England has advised at least 74 local planning authorities (LPAs) that, where protected sites are in unfavourable condition due to excess nutrients, projects and plans should only go ahead if they will not cause additional pollution to sites. LPAs can demonstrate this through 'nutrient neutrality'. If the nutrient load created through additional wastewater from a development is mitigated, the LPA may permit a new development to go ahead. However, this will require the creation of new wetlands to strip nutrients from water, or the creation of buffer zones that are allowed to revert to nature (Defra *et al.*, 2022b).

Since December 2022, Natural England has been approaching landowners in a targeted way to invite them to offer their land as potential sites for nutrient mitigation. These sites will start to provide the mitigation needed by LPAs and developers, and the programme will be expanded across the country to facilitate the building of thousands of new homes, as well as making a major contribution to nature recovery (Defra *et al.*, 2022b).

6.11.4.1 Funding

To enable developments to continue, the government has offered £100 000 to each catchment area to assist local authorities with meeting the Natural England requirements. The funding is used to employ catchment officers to deal with mitigation measures. More information about funding support is available on the government website (Defra, 2022).

6.11.4.2 Nutrient mitigation scheme and planning permission

The government has also worked with Natural England to produce a new nutrient mitigation scheme. This will be delivered by Natural England by establishing an 'Accelerator Unit', with the support of Defra, DLUHC, the Environment Agency and Homes England (Defra *et al.*, 2022b).

The scheme will require developers to purchase 'nutrient credits' to discharge their requirements to provide mitigation, if they wish to develop in areas where are there are nutrient pollution concerns, such as protected wildlife areas, wetlands and woodlands. Natural England will provide accreditation for mitigation under the scheme.

LPAs will then grant planning permission for developments which have secured the necessary nutrient credits, with priority given to smaller builders, which have limited resources. This will ensure developers have a streamlined way to mitigate nutrient pollution, allowing planned building to continue and the creation of new habitats across the country.

The scheme also provides for Natural England to work in partnership with environmental groups and other privately led nutrient mitigation schemes to mitigate the impacts of unavoidable nutrient pollution through the creation of new wetlands and woodlands, and new habitats (Defra *et al.*, 2022b).

References

Parliamentary bills

Levelling Up and Regeneration bill

Statutes

Environment Act 2021
Land Drainage Act 1991
Marine and Coastal Access Act 2009
Reservoirs Act 1975
Water Act 1989
Water Act 2003
Water Act 2014
Water Industry Act 1991
Water Resources Act 1991

Regulations and directions

Environmental Permitting (England and Wales) Regulations 2016
Floods and Water (Amendment etc.) (EU Exit) Regulations 2019
The Conservation of Habitats and Species Regulations 2017
The Environmental Damage (Prevention and Remediation) (England) Regulations 2015
The Environmental Damage (Prevention and Remediation) (Wales) Regulations 2009
The Flood Plan (Reservoirs Emergency Planning) Direction 2021
The Water Environment (Water Framework Directive) (England and Wales) Regulations 2003
The Water Environment (Water Framework Directive) (England and Wales) Regulations 2017
The Water Environment (Water Framework Directive) Regulations (Northern Ireland) 2017
The Water Supply (Water Fittings) Regulations 1999

Directives

The Habitats Directive 92/43/EEC

Case law

Harris & Anor v The Environment Agency [2022] EWHC 2606

Journals

Elmes V and Coleman N (2021) Interactions of Cd^{2+}, Co^{2+} and MoO_4^{2-} ions with crushed concrete fines. *Journal of Composites Science* **5(2)**: 42. 10.3390/jcs5020042.

Yoo J, Shin H and Ji S (2017) An eco-friendly neutralization process by carbon mineralization for Ca-rich alkaline wastewater generated from concrete sludge. *Metals – Open Access Metallurgy Journal* **7(9)**: 371. 10.3390/met7090371.

Websites

Coulter RD (2019) Top sources of water pollution on construction sites and how to treat it. *Construction UK*, 14 August. https://constructionmaguk.co.uk/top-sources-of-water-pollution-on-construction-sites-and-how-to-treat-it (accessed 09/03/2023).

Coulton R (2016) How to treat concrete washwater. *Construction News*, 14 April. https://www.constructionnews.co.uk/special-reports/how-to-treat-concrete-washwater-14-04-2016 (accessed 09/03/2023).

Defra (Department for Environment, Food and Rural Affairs) (2021a) Water abstraction plan 2017. https://www.gov.uk/government/publications/water-abstraction-plan-2017 (accessed 09/03/2023).

Defra (2021b) Reservoir owner and undertaker responsibilities: on-site emergency flood plans. https://www.gov.uk/government/publications/reservoir-emergencies-on-site-plan/reservoir-owner-and-undertaker-responsibilities-on-site-emergency-flood-plans (accessed 09/03/2023).

Defra (2021c) Water abstraction plan 2017. https://www.gov.uk/government/publications/water-abstraction-plan-2017 (accessed 09/03/2023).

Defra (2021d) *Changes to the Regulatory Framework for Abstraction and Impounding Licensing in England.* https://consult.defra.gov.uk/water/abstraction-impounding-epr-consultation/supporting_documents/Supplementary%20Document%20%20AI%20move%20into%20the%20EPR.pdf (accessed 09/03/2023).

Defra (2022) Nutrient pollution: reducing the impact on protected sites. https://www.gov.uk/government/publications/nutrient-pollution-reducing-the-impact-on-protected-sites/nutrient-pollution-reducing-the-impact-on-protected-sites (accessed 09/03/2023).

Defra and Environment Agency (2022) Review individual flood risk assessments: standing advice for local planning authorities. https://www.gov.uk/guidance/flood-risk-assessment-local-planning-authorities (accessed 09/03/2023).

Defra, Forestry Commission, Environment Agency and Natural England (2019) First review of 25 Year Environment Plan published. Press release. https://www.gov.uk/government/news/first-review-of-25-year-environment-plan-published (accessed 09/03/2023).

Defra, DLUHC (Department for Levelling Up, Housing and Communities) and Natural England (2022a) Government sets out plan to reduce water pollution. https://www.gov.uk/government/news/government-sets-out-plan-to-reduce-water-pollution (accessed 09/03/2023).

Defra, DLUHC and Natural England (2022b) Government sets out plan to reduce water pollution. https://www.gov.uk/government/news/government-sets-out-plan-to-reduce-water-pollution (accessed 09/03/2023).

DLUHC (Department for Levelling Up, Housing and Communities) (2022) Levelling-up and Regeneration Bill: reforms to national planning policy. https://www.gov.uk/government/consultations/levelling-up-and-regeneration-bill-reforms-to-national-planning-policy/levelling-up-and-regeneration-bill-reforms-to-national-planning-policy (accessed 09/03/2023).

Edgar A (2018) Harron Homes fined £120,000 for water pollution. IEMA. https://www.iema.net/articles/harron-homes-fined-120-000-for-water-pollution (accessed 09/03/2023).

Environment Agency (2015) Standard Rules SR2015 No. 30 – Temporary Diversion of a Main River. https://assets.publishing.service.gov.uk/government/uploads/system/uploads/attachment_data/file/823466/Standard-Rule-2015-No-30.pdf (accessed 09/03/2023).

Environment Agency (2017) Harron Homes fined £120,000 over construction pollution. https://www.gov.uk/government/news/harron-homes-fined-120000-over-construction-pollution (accessed 09/03/2023).

Environment Agency (2019) *Polycyclic Aromatic Hydrocarbons (PAHs): Sources, Pathways and Environmental Data.* https://consult.environment-agency.gov.uk/environment-and-business/challenges-and-choices/user_uploads/polycyclic-aromatic-hydrocarbons-rbmp-2021.pdf (accessed 09/03/2023).

Environment Agency (2020) Treating and using water that contains concrete and silt at construction sites: RPS 235. https://www.gov.uk/government/publications/treating-and-using-water-that-contains-concrete-and-silt-at-construction-sites-rps-235/treating-and-using-water-that-contains-concrete-and-silt-at-construction-sites-rps-235 (accessed 09/03/2023).

Environment Agency (2021) General binding rules: small sewage discharge to a surface water. https://www.gov.uk/guidance/general-binding-rules-small-sewage-discharge-to-a-surface-water (accessed 09/03/2023).

Environment Agency (2022a) Apply for a water abstraction or impounding licence. https://www.gov.uk/guidance/water-management-apply-for-a-water-abstraction-or-impoundment-licence (accessed 09/03/2023).

Environment Agency (2022b) Environmental permits and abstraction licences: tables of charges. https://www.gov.uk/government/publications/environmental-permits-and-abstraction-licences-tables-of-charges (accessed 09/03/2023).

Environment Agency (2023) Temporary dewatering from excavations to surface water: RPS 261. https://www.gov.uk/government/publications/temporary-dewatering-from-excavations-to-surface-water (accessed 09/03/2023).

Environment Agency and Defra (2018) Groundwater protection position statements. https://www.gov.uk/government/publications/groundwater-protection-position-statements (accessed 09/03/2023).

Environment Agency and Defra (2022a) Discharges to surface water and groundwater: environmental permits. https://www.gov.uk/guidance/discharges-to-surface-water-and-groundwater-environmental-permits (accessed 09/03/2023).

Environment Agency and Defra (2022b) Discharges to surface water and groundwater: environmental permits – Discharges in sewered areas. https://www.gov.uk/guidance/discharges-to-surface-water-and-groundwater-environmental-permits#discharges-in-sewered-areas (accessed 09/03/2023).

Hughes M (2022) Creating the new homes and the healthy natural environment we need. Natural England blog. https://naturalengland.blog.gov.uk/2022/03/18/creating-the-new-homes-and-the-healthy-natural-environment-we-need (accessed 09/03/2023).

Natural England (2022) Letter from Natural England, 16 March 2022 – Advice for development proposals with the potential to affect water quality resulting in adverse nutrient impacts on habitats sites. https://www.cornwall.gov.uk/media/gbpk3k1y/ne-water-quality-and-nutrient-neutrality-advice-16_03_2022-issue-1-final.pdf (accessed 09/03/2023).

Natural England, Defra and DLUHC (2022) *Nutrient Neutrality. A Summary Guide*. https://www.cornwall.gov.uk/media/1hmjm0bm/nutrient-neutrality-a-summary-guide.pdf (accessed 09/03/2023).

UK Parliament (2022) Levelling-up and Regeneration bill. https://bills.parliament.uk/bills/3155/news (accessed 09/03/2023).

WRAS (Water Regulations Approval Scheme) (2023) https://www.wrasapprovals.co.uk (accessed 09/03/2023).

Baker F and Charlson J
ISBN 978-0-7277-6645-8
https://doi.org/10.1680/elsc.66458.169

Chapter 7
Site hazards and nuisances

Jennifer Charlson

7.1. Introduction

This chapter identifies risks which contractors should consider before work commences on site. Contractors can then take steps to minimise environmental impact and avoid confrontation with regulators, and the risk of prosecution, criminal fines, actions for damages and injunctions.

Contractors should be aware that, under the Occupiers' Liability Act 1957 (lawful visitors) and Occupiers' Liability Act 1984 (unlawful visitors, i.e. trespassers), they owe a duty of care to visitors and trespassers on their sites. Cases establishing a duty to child trespassers are reported. Subsequently, a statutory qualified duty to a trespasser was introduced.

The Control of Substances Hazardous to Health Regulations 2002 (COSHH regulations) provide a framework to protect employees against health risks from hazardous substances. The COSHH regulations are applicable as contractors bring to site construction products which, if not handled appropriately, can cause harm. Managing and working with asbestos in non-domestic buildings is regulated under the Control of Asbestos Regulations 2012.

The Hazardous Waste (England and Wales) Regulations 2005 set out the regime for the control and tracking of hazardous waste in England and Wales.

Part III of the Environmental Protection Act 1990 concerns 'statutory nuisance' in relation to noise, dust and odour. Local authorities are under a duty to inspect their areas to detect statutory nuisances and take enforcement action when they detect or anticipate a nuisance being caused. Pursuant to the Control of Pollution Act 1974, a local authority can serve a notice which imposes constraints on the operations to minimise noise from a construction site. In England and Wales, concentrations of key pollutants in outdoor air are regulated by the Air Quality Standards Regulations 2010.

Nuisance occurs when some condition or activity interferes with the use or enjoyment of land. Where the nuisance affects the rights of a number of persons, it may be classified as a public nuisance. Private nuisance is defined as an unlawful interference by a person with another's use and enjoyment of his land. The *Rylands v Fletcher* cause of action (see Section 7.9.3) relates to strict liability for foreseeable damage caused by escapes occasioned by the non-natural use of land.

It is possible for liability in negligence to arise in relation to environmental problems. However, the duty of care in negligence is restricted to cases of physical damage and injury rather than pure economic loss. An action for trespass to land can be brought by a person entitled to possession of the land against a person whose unlawful act causes direct physical interference with the land.

7.2. Occupiers' Liability Acts 1957 and 1984

Contractors should be aware that, under the Occupiers' Liability Act 1957 (lawful visitors) and Occupiers' Liability Act 1984 (unlawful visitors, i.e. trespassers), they owe a duty of care to visitors and trespassers on their sites. The duty owed to a visitor is defined by section 2(2) of the 1957 Act, which imposes upon an occupier

> A duty to take such care as in all the circumstances is reasonable to see that the visitor will be reasonably safe in using the premises for the purposes for which he is invited or permitted by the occupier to be there.

7.2.1 Liability for child trespassers

Two cases that established liability for child trespassers are reported.

Box 7.1 *British Railways Board v Herrington* [1972] AC 877

This case considered the liability of the British Railways Board for the electrocution of a child who trespassed on a railway line. Children regularly used a gap in a fence, which surrounded a railway line, as a short cut to a park. The defendant was aware of the disrepair of the fence but had failed to act.

The House of Lords, departing from its previous authority, held that the defendant owed a duty of common humanity to trespassers. Lord Pearson reasoned that

> There is considerably more need than there used to be for occupiers to take reasonable steps with a view to deterring persons, especially children, from trespassing in places that are dangerous for them.

Box 7.2 *Pannett v McGuinness and Co Ltd* [1972] 3 All ER 137

Demolition contractors McGuinness and Co Ltd were burning rubbish in a busy urban area. As they were aware that the fire would attract children, workmen were employed to deter them. However, while the fire was still burning, the workmen left and the plaintiff was injured.

As the activity was hazardous and the likelihood of trespass was great, due to the obvious attraction to children, the Court of Appeal found the defendant liable.

7.2.2 Qualified duty to a trespasser

Subsequent to the cases outlined above, a qualified duty to a trespasser was introduced by the Occupiers' Liability Act 1984, limited by section 1(4) to

> the duty … to take such care as is reasonable in all the circumstances of the case to see that he does not suffer injury on the premises by reason of the danger concerned

and arising only under section 1(3) of that Act, where

- the occupier knows of the danger or has constructive knowledge of it
- the occupier knows, or has constructive knowledge, that the trespasser is in the vicinity or may come into the vicinity of the danger
- the danger is one which, in all circumstances, the occupier can reasonably be expected to offer protection against.

Therefore, as construction sites are enticing to children, contractors should guard against trespassing. For example, the site perimeter should be secured and potential hazards stored or enclosed. The Health and Safety Executive (HSE) has published guidance about protecting the public, including vulnerable groups (HSE, 2023a).

7.3. Control of Substances Hazardous to Health Regulations 2002

The Control of Substances Hazardous to Health Regulations 2002 (COSHH regulations) provide a framework to protect employees against health risks from hazardous substances. The regulations require employers to

- assess the risks
- decide on the required precautions
- prevent or control the risks
- ensure that control measures are used and maintained
- monitor exposure and implement health surveillance, when necessary
- inform, instruct and train employees about the risks and precautions.

The COSHH regulations are applicable as contractors bring to site construction products which, if not handled appropriately, can cause harm. Such substances include paints, solvents, cement and fuels (used for plant and machinery) which can be harmful for personnel and the environment.

The HSE provides practical advice and guidance on the COSHH regulations, including advice on completing COSHH assessments (HSE, 2023b).

7.4. Control of Asbestos Regulations 2012

Managing and working with asbestos in non-domestic buildings is regulated under the Control of Asbestos Regulations 2012.

Regulation 4 obliges the 'duty holder' to manage the risk from asbestos in non-domestic premises, so as to ensure that workers are no longer knowingly exposed to any potential or real risk. Duty holders must

- take reasonable steps to identify asbestos within their premises, the amount present and its condition
- presume materials contain asbestos unless there is evidence that they do not

- complete a written record of the location of the asbestos or presumed asbestos
- carry out a risk assessment
- prepare a plan setting out how any risk will be managed
- implement the plan
- pass on information about asbestos containing materials to those who need it
- monitor and review all the above arrangements periodically.

Non-compliance with regulation 4 is a criminal offence, punishable with a fine of an unlimited amount and/or imprisonment for up to two years.

The HSE (2013) has published the *Control of Asbestos Regulations 2012, the Approved Code of Practice and Guidance* for use by employers. It covers work which disturbs, or is likely to disturb, asbestos, and asbestos sampling and laboratory analysis.

7.5. Hazardous waste regulations

The Hazardous Waste (England and Wales) Regulations 2005, which came into force in July 2005, set out the regime for the control and tracking of hazardous waste in England and Wales. The regulations introduced a process of registration of hazardous waste producers and a new system for recording the movement of waste.

7.5.1 Waste

A waste is defined as hazardous if it is included in the List of Wastes (England) Regulations 2005. Examples of hazardous waste include

- asbestos
- chemicals, such as brake fluid or printer toner
- batteries
- solvents
- pesticides
- oils (except edible ones), such as car oil
- equipment containing ozone-depleting substances, such as fridges
- hazardous waste containers.

7.5.2 Producer registration

All industrial and commercial premises producing more than 500 kilograms of hazardous waste have to notify their existence to the Environment Agency. In addition

- no hazardous waste can be collected from any unregistered site
- waste producers who do not register their premises will be committing an offence
- waste contractors who move waste from a non-registered site will be committing an offence
- waste producers will need to provide proof to waste contractors (via a unique code number) that they are registered
- registrations must be renewed annually.

7.5.3 Movement of hazardous waste

The movement of wastes is controlled by a documentation system that has to be completed whenever waste is removed from premises. From the waste producer's perspective, the most important form is the consignment note. This has to be completed before waste can be removed, and detailed information has to be provided. This information must include a description of the waste, detailing

- the process giving rise to the waste
- the quantity
- the chemical (and/or biological) components and their concentrations
- the hazard codes
- the container type, size and number.

In addition, the consignment note must

- identify where the waste is going
- have a unique code number.

A copy of the consignment notes and the quarterly return in a register, for a minimum of three years, must be made, and be available to the Environment Agency on request.

Guidance on hazardous waste, including producer or holder, carrier and consignee requirements is available on the government website (HMG, 2023).

7.6. Environmental Protection Act 1990 – statutory nuisance

Part III of the Environmental Protection Act 1990 concerns 'statutory nuisance' in relation to noise, dust and odour. Pursuant to section 79, local authorities are under a duty to inspect their areas, from time to time, to detect statutory nuisances and take enforcement action when they detect or anticipate a nuisance being caused. As a public body is responsible for enforcement, statutory nuisance represents more of a threat than a common-law action, which usually requires the claimant to finance the cost of enforcement.

7.6.1 Statutory nuisance

Matters which might give rise to statutory nuisance are defined in section 79(1) of the Environmental Protection Act 1990, and include

- premises in such a state as to be prejudicial to health or a nuisance
- smoke, fumes or gases emitted from premises so as to be prejudicial to health or a nuisance
- dust, steam, smell or other effluvia arising on industrial, trade or business premises and being prejudicial to health or a nuisance
- any accumulation or deposit which is prejudicial to health or a nuisance
- noise emitted from or caused by a vehicle, machinery or equipment in a street that is prejudicial to health or a nuisance.

'Prejudicial to health' is defined at section 79(7) as meaning injurious, or likely to cause injury, to health. 'Nuisance' is regarded as having the same meaning as under the common law.

7.6.2 Abatement notice

Section 80 of the Environmental Protection Act 1990 provides that, where a local authority is satisfied that a statutory nuisance exists, or is likely to occur or recur, it is under a duty to serve an abatement notice on the person responsible for the nuisance, or, if that person cannot be found, on the owner or occupier of the site from which the statutory nuisance emanates. A statutory notice differs from the tort of nuisance in that there is no requirement for the notice to be supported by evidence that a particular neighbour has suffered unreasonable interference with the enjoyment of his property.

However, the wording of the notice must be clear and precise because non-compliance with the terms of the notice can lead to criminal sanctions. The person served with the notice can appeal to the magistrates' court within 21 days of service.

7.6.3 Complaint to the local magistrates' court

As environmental health departments are often overworked and under-resourced, the statutory nuisance provisions tend not to be used in the proactive way suggested by section 79(1) of the Environmental Protection Act 1990, which imposes a duty on local authorities to inspect their areas from time to time to detect statutory nuisances. In practice, what more often happens is that aggrieved individuals notify the authority about potential problems. The relevant authority then has a duty to take such steps as are reasonably practicable to investigate.

Where the local authority is not inclined to act, section 82 allows a complaint to be made to the local magistrates' court by any person who is aggrieved by the existence of a statutory nuisance. Where the complainant can convince the magistrates that there is an existing or recurring nuisance, the magistrates have a duty to serve an abatement notice requiring the defendant to abate the nuisance within a specified time, or to carry out such works so as to prevent the recurrence of the nuisance.

7.6.4 Defences for non-compliance with an abatement notice

There are various defences which can be deployed for non-compliance with an abatement notice, but where there is no applicable defence the person on whom the notice is served will be criminally liable for non-compliance with the notice. Section 80(7) of the Environmental Protection Act 1990 offers a defence where the best practicable means have been to avert the nuisance, and there is a special defence in relation to noise and nuisances on construction sites pursuant to section 80(9).

7.6.5 DEFRA guidance

The Department for Environment, Food and Rural Affairs (Defra) has published guidance on

- what counts as a statutory nuisance and how councils can deal with complaints by issuing an abatement notice (Defra, 2015a)
- how councils deal with complaints about noise (including from construction works) (Defra, 2017a).

Defra (2017b) has also published guidance setting out the government's interpretation of the relationship between environmental permitting and local authorities' statutory nuisance duties in England and Wales.

7.7. Control of Pollution Act 1974 – noise

Section 60 of the Control of Pollution Act 1974 applies to a wide range of engineering construction activities from the construction and demolition of buildings, structures or roads to dredging work. Pursuant to this section, a local authority can serve a notice which imposes constraints on the operations to minimise noise from a construction site. The notice may specify

- hours of working
- types of plant and machinery which may or may not be used on site
- levels of noise which may be emitted from the site during specified hours.

A person served with a section 60 notice may appeal, within 21 days, to a magistrates' court. It is an offence to fail to comply with a section 60 notice.

7.7.1 Consent application

Under section 61 of the Control of Pollution Act 1974, a contractor can apply for a consent in advance of commencing work. The notice represents an agreed compromise, about the construction operations, between the contractor and the local authority to minimise noise. The notice also acts as a defence to a section 60 notice.

However, many contractors decide not to apply for a section 61 consent because it may draw attention to the site and its potential for noise. Furthermore, it is an offence to contravene any of the consent conditions. Therefore, many contractors prefer to work without a consent and accept the risk that the local authority may serve a section 60 notice.

7.8. Air pollution

In England and Wales, concentrations of key pollutants in outdoor air are regulated by the Air Quality Standards Regulations 2010 and the Air Quality Standards (Wales) Regulations 2010.

7.8.1 Exposure limits

The air quality standards regulations (see above) seek to control human exposure to pollutants in outdoor air to protect human health and the environment by requiring concentrations to be within specified limit values. The regulations set

- legally binding limits for concentrations in outdoor air of major air pollutants that impact public health: sulfur dioxide, nitrogen oxides, particulate matter (as PM_{10} and $PM_{2.5}$), lead, benzene, carbon monoxide and ozone
- targets for levels in outdoor air for four elements: cadmium, arsenic, nickel and mercury, together with polycyclic aromatic hydrocarbons (PAHs).

Separate legislation exists for emissions of air pollutants. The National Emission Ceilings Regulations 2018 set national (UK-wide) emission limits or 'ceilings' for sulfur dioxide, oxides of nitrogen and ammonia non-methane volatile organic compounds in 2010, 2020 and 2030, and for $PM_{2.5}$ in 2020 and 2030.

7.8.2 National and local authorities

The responsibility for meeting the exposure limits in England lies with the Secretary of State for Environment, Food and Rural Affairs, and in Wales it is devolved to the National Assembly for Wales. Defra coordinates assessment and air quality plans for the UK as a whole. It also publishes annual estimates of UK emissions of particulate matter (PM_{10} and $PM_{2.5}$), nitrogen oxides, ammonia, non-methane volatile organic compounds and sulfur dioxide (Defra, 2023).

The Environment Act 1995 requires the government to produce a national air quality strategy (AQS) for the UK, setting out air quality standards, objectives and measures for improving ambient air quality. The last comprehensive review of the AQS was published in 2007, with a review yielding some minor changes published in 2011. The AQS sets out the UK's air quality objectives and recognises that action at national, regional and local level may be needed, depending on the scale and nature of the air quality problem.

Under the Environment Act 2021, the Secretary of State must review the AQS for England at least every five years, with a commitment for an initial review within 12 months of the measures coming into force. The first review will be published in 2023.

Part IV of the Environment Act 1995 requires local authorities in the UK to review air quality in their area and designate air quality management areas if improvements are necessary. Where an air quality management area is designated, local authorities are also required to work towards the AQS's objectives. An air quality action plan describing the pollution reduction measures must then be put in place.

7.9. Public and private nuisance

Nuisance occurs when some condition or activity interferes with the use or enjoyment of land.

7.9.1 Public nuisance

Where the nuisance affects the rights of a number of persons, it may be classified as a public nuisance.

Box 7.3 *Attorney General v PYA Quarries* [1957] 2 QB 169

This case identified public nuisance as materially affecting the reasonable comfort and convenience of a class of Her Majesty's subjects. The defendants operated a quarry using a blasting technique that generated dust, noise and vibration.

A further example of public nuisance is obstruction of the public highway. Public nuisance is primarily a crime, but it may be possible for a person affected to sue in tort.

The HSE (2023c) has provided guidance, in the format of FAQs, on how to resolve typical construction public nuisance scenarios, including dust, noise or smoke from demolition sites, and mud or construction debris on the public road from vehicles leaving a demolition or construction site.

7.9.2 Private nuisance

Private nuisance is defined as an unlawful interference by a person with another's use and enjoyment of his land.

Box 7.4 *Andreae v Selfridge & Co Ltd* [1937] 3 All ER 255

The claimant, a hotel owner, recovered damages from Selfridge, who had created excessive noise and dust during demolition work in developing part of the same London block as occupied by the hotel.

Box 7.5 *Video London Sound Studios Ltd v Asticus (GMS) Ltd and Keltbray Demolition Ltd* [2001] All ER (D) 168

In this more recent case, the specialist sound engineers succeeded in their private nuisance claim against the demolition contractor. The demolition works caused debris and dust to fall from the plaintiff's chimney into its basement, resulting in damage to sensitive electronic equipment.

Box 7.6 *Hunter and Others v Canary Wharf Ltd and the London Docklands Development Corporation* [1997] AC 655, HL

The right of private action is restricted, so that only those who have suffered direct foreseeable interference with their physical wellbeing, their personal property or their use or enjoyment of land can sue. In this case, the House of Lords stated that only a person with an interest in land could sue in nuisance.

7.9.3 The rule in *Rylands v Fletcher*

This cause of action relates to strict liability for foreseeable damage caused by escapes occasioned by the non-natural use of land.

Box 7.7 *Rylands v Fletcher* [1868] UKHL 1

This 19th century House of Lords case introduced the rule that the person who for his own purposes brings on his lands and collects and keeps there anything likely to do mischief if it escapes, must keep it in at his peril, and, if he does not do so, is *prima facie* answerable for all the damage which is the natural consequence of its escape.

The case concerned a reservoir created by the owner of the land from which water escaped, flooding a mine on neighbouring land. The owner of the reservoir was held to be liable for the damage caused.

7.9.3.1 *Rylands v Fletcher* cases

Box 7.8 *Cambridge Water Co Ltd v Eastern Counties Leather plc* [1994] 1 All ER 53

This House of Lords case established the principle that claims under *Rylands v Fletcher* must include a requirement that damage be foreseeable.

The Cambridge Water Company had operated a borehole which was contaminated with perchloroethene that had originated in a tannery owned by Eastern Counties Leather. However, the court found that Eastern Counties Leather was not liable to pay damages as the environmental harm had not been reasonably foreseeable.

Box 7.9 *Transco plc v Stockport Metropolitan Borough Council* [2003] UKHL 61

In this later House of Lords case, it was held that, because quantities of water from an ordinary pipe is not dangerous or unnatural, the council was not liable. The ground under a gas pipe had washed away when the council's water pipe leaked.

7.10. Negligence

It is possible for liability in negligence to arise in relation to environmental problems. In order to establish liability, there has to be

- a duty of care owed to the claimant by the defendant, and
- a breach of that duty, and
- damage caused as a result of the breach.

In a construction context, negligence claims are commonly made against the design team and/or design-and-build contractor. The distinction between a construction professional and a design-and-build contractor's standard of design responsibility is detailed in Charlson (2019).

7.10.1 No recovery of pure economic loss

Box 7.10 *D & F Estates Ltd v Church Commissioners for England and Wales* [1989] AC 177

This was a landmark House of Lords judgement which restricted the duty of care in negligence to cases of physical damage and injury rather than pure economic loss.

The Church Commissioners owned a block of flats where the plastering work, which was subcontracted, was defective. As there was no direct contractual relationship between the plaintiff and the defendants, an action was brought in tort.

Box 7.11 *Thomas & Anor v Taylor Wimpey Developments Ltd* [2019] EWHC 1134

The claimants sought damages to rectify alleged defective log retaining walls at the rear of their back gardens. The court upheld the long-standing principle that the builder did not owe a duty of care to the home owners in tort in respect of pure economic loss.

7.11. Trespass

An action for trespass to land can be brought by a person entitled to possession of the land against a person whose unlawful act causes direct physical interference with the land.

Box 7.12 *Jones v Llanrwst Urban District Council* [1911] 1 Ch 393

A landowner complained that Llanrwst Urban District Council was discharging untreated sewage into a river, causing solid waste to be deposited on his land. The judge held that this was trespass to the plaintiff's land. Nevertheless, trespass is not frequently used as a cause of action for pollution incidents.

However, trespass remains relevant to the construction industry.

Box 7.13 *Anchor Brewhouse Developments Ltd and Ors v Berkley House (Docklands) Developments Ltd* [1987] 38 BLR 82

In this case, the court granted an injunction, even where no damage was caused, to prevent oversailing by tower cranes.

Box 7.14 *Wollerton and Wilson Ltd v Richard Costain Ltd* [1970] 1 WLR 411

In this case, the court suspended an injunction when the landowner had refused a reasonable offer of payment and insurance.

The Institute of Party Wall Surveyors' (IPWS, 2023) advice is to approach all landowners whose property may be oversailed to seek their permission to do so, which should be documented in a crane oversail licence.

References

Statutes

Control of Pollution Act 1974
Environment Act 1995
Environment Act 2021

Environmental Protection Act 1990
Occupiers' Liability Act 1957
Occupiers' Liability Act 1984

Regulations

Air Quality Standards Regulations 2010
Air Quality Standards (Wales) Regulations 2010
Control of Asbestos Regulations 2012
The Control of Substances Hazardous to Health Regulations 2002
The Hazardous Waste (England and Wales) Regulations 2005
The List of Wastes (England) Regulations 2005
The National Emission Ceilings Regulations 2018

Case law

Anchor Brewhouse Developments Ltd and Ors v Berkley House (Docklands) Developments Ltd [1987] 38 BLR 82
Andreae v Selfridge & Co Ltd [1937] 3 All ER 255
Attorney General v PYA Quarries [1957] 2 QB 169
British Railways Board v Herrington [1972] AC 877
Cambridge Water Co Ltd v Eastern Counties Leather plc [1994] 1 All ER 53
D & F Estates Ltd v Church Commissioners for England and Wales [1989] AC 177
Hunter and Others v Canary Wharf Ltd and the London Docklands Development Corporation [1997] AC 655, HL
Jones v Llanrwst Urban District Council [1911] 1 Ch 393
Pannett v McGuinness and Co Ltd [1972] 3 All ER 137
Rylands v Fletcher [1868] UKHL 1
Thomas & Anor v Taylor Wimpey Developments Ltd [2019] EWHC 1134
Transco plc v Stockport Metropolitan Borough Council [2003] UKHL 61
Video London Sound Studios Ltd v Asticus (GMS) Ltd and Keltbray Demolition Ltd [2001] All ER (D) 168
Wollerton and Wilson Ltd v Richard Costain Ltd [1970] 1 WLR 411

Journals

Charlson J (2019) Briefing: Interpreting contractors' mandated standard of design. *Proceedings of the Institution of Civil Engineers – Management, Procurement and Law* **172(4)**: 142–145. 10.1680/jmapl.19.00007.

Websites

Defra (Department for Environment, Food and Rural Affairs) (2015) Statutory nuisances: how councils deal with complaints. https://www.gov.uk/guidance/statutory-nuisances-how-councils-deal-with-complaints (accessed 09/03/2023).
Defra (2017a) Noise nuisances: how councils deal with complaints. https://www.gov.uk/guidance/noise-nuisances-how-councils-deal-with-complaints (accessed 09/03/2023).
Defra (2017b) Environmental permitting and statutory nuisance. https://www.gov.uk/government/publications/environmental-permitting-guidance-statutory-nuisance (accessed 09/03/2023).

Defra (2023) Emissions of air pollutants. https://www.gov.uk/government/statistics/emissions-of-air-pollutants (accessed 09/03/2023).

HMG (His Majesty's Government) (2023) Hazardous waste. https://www.gov.uk/dispose-hazardous-waste (accessed 09/03/2023).

HSE (Health and Safety Executive) (2013) *Managing and Working with Asbestos. Control of Asbestos Regulations 2012. Approved Code of Practice and Guidance.* https://www.hse.gov.uk/pubns/books/l143.htm (accessed 09/03/2023).

HSE (2023a) Protecting the public. https://www.hse.gov.uk/construction/safetytopics/publicprotection.htm (accessed 09/03/2023).

HSE (2023b) Control of Substances Hazardous to Health (COSHH). https://www.hse.gov.uk/coshh (accessed 09/03/2023).

HSE (2023c) Construction general: Public nuisance. https://www.hse.gov.uk/construction/faq-publicnuisance.htm (accessed 09/03/2023).

IPWS (Institute of Party Wall Surveyors) (2023) Oversailing: what is it and why is it a concern for developers? https://www.ipws.co.uk/advice/case-law/17-oversailing-what-is-it-and-why-is-it-a-concern-for-developers (accessed 09/03/2023).

Baker F and Charlson J
ISBN 978-0-7277-6645-8
https://doi.org/10.1680/elsc.66458.183

Chapter 8
Insurance

Jennifer Charlson

8.1. Introduction

This chapter explains insurance in the context of the construction industry. The decision to purchase insurance policies is examined. An uninsured excess, the maximum amount an insurer will pay and an insured's obligations are explained. 'Claims-made' and 'occurrence' basis claims are distinguished.

Specific insurance policies are then outlined. Contractors' all risks insurance (CAR) primarily refers to material damage cover for the contract works. Public liability insurance will provide indemnity for liability for damages arising from accidental injury to third parties (not employees) or accidental damage to third-party property as a consequence of the project. Cases involving claims for environmental costs against public liability insurance are reported. In Great Britain, employer's liability insurance is mandated by the Employers' Liability (Compulsory Insurance) Act 1969.

Professional indemnity (PI) insurance covers awards of damages, costs or settlements resulting from a claim made during the policy period for any act, error or omission arising out of the conduct of the business. A Court of Appeal case about a professional negligence claim against a surveyor is reported. A fitness-for-purpose standard accepted by contractors will not be back-to-back with a designer who will only accept a reasonable skill and care responsibility. PI insurance policies will cover a failure to exercise reasonable skill and care but usually exclude protection for fitness-for-purpose obligations.

A structural warranty is an insurance policy designed to protect against defects in new buildings, for a defined period known as the 'policy term'. Various Joint Contracts Tribunal (JCT) contract conditions mandate non-negligent liability insurance to protect the client against liability for damage to adjacent or surrounding property.

Directors' and officers' (D&O) liability insurance cover is for awards of damages, costs or settlements for which a director or officer is liable, resulting for a claim made against them during the policy period for any act, error or omission in their capacity as a director or officer of the company.

Environmental insurance can be defined as protection against an insured's exposure to the liabilities arising from the ownership and/or development of brownfield land and any consequential loss and damages. The availability of environmental insurance, exposure to environmental claims and environmental insurance options are examined. Generally, in circumstances where a polluting incident occurs, the polluter is not obliged to report the incident to the regulatory authorities.

Finally, The Chancery Lane Project insurance-focused climate clauses are explored.

8.1.1 Insurance policies

The decision to purchase insurance policies is informed by many factors, including risk assessment, contractual or professional body requirements, and statutory obligation. An insurance policy is one of 'good faith', requiring the insured to disclose relevant information to the insurer for assessment and pricing of risk.

All insurance policies have an uninsured excess (or deductible) that is payable by the insured in the event of a claim. There is also a limit on the maximum amount the insurer will pay on any single claim, and there may be an overall maximum amount the insurer will pay in a single policy.

Under terms of cover, the insured is required to promptly notify claims and circumstances that may give rise to claims. Insurance policies apply either to claims made against the insured during the period of cover (i.e. 'claims-made' basis) or incidents occurring during the period of cover (i.e. an 'occurrence' basis).

8.2. All-risks insurance

Contractors' all-risks insurance (CAR) primarily refers to material damage cover of the contract works. A CAR policy responds when the works being constructed are damaged by an insured peril and require replacing and/or repairing.

The project contract will normally stipulate whether the contractor or employer is to provide the cover, which should be in their joint names. The policy will typically respond to any physical loss or damage, for example, protection against fire, earthquake, malicious damage and other accidental damage. Alternatively, policies can be issued covering loss or damage by specified perils, for example, fire, flood and storm. Exclusions are also likely to apply to the policy.

8.3. Public liability insurance

Public liability insurance will provide indemnity for liability for damages arising from accidental injury to third parties (not employees) or accidental damage to third-party property as a consequence of the project. However, environmental claims (e.g. asbestos and gradual pollution) may be excluded.

Cover is frequently required by contract insurance clauses. An annual policy will have a maximum amount payable in the event of one claim or series of claims arising from one occurrence. A project policy may be arranged that covers specified project participants, such as the employer, contractor and subcontractors.

8.3.1 Environmental costs claims against public liability insurance

There have been unsuccessful attempts to recover environmental costs from public liability policies.

Box 8.1 *Bartoline Limited v Royal & Sun Alliance Insurance plc* [2006] EWHC 3598 (QB)

In this case, the public liability policy was found not to cover clean-up costs incurred by the Environment Agency exercising its statutory powers or the insured's own costs in complying with Water Resources Act 1991 notices.

Bartoline manufactured adhesives and packed hydrocarbons such as white spirit and turpentine. Following a fire at its premises, chemicals and fire-fighting foam heavily contaminated two nearby watercourses.

Box 8.2 *Yorkshire Water Services Ltd v Sun Alliance & London Insurance plc and Others* [1997] 2 Lloyd's Rep. 21

The judge in *Bartoline Limited v Royal & Sun Alliance Insurance plc* (see Box 8.1) followed the Court of Appeal's decision in this case, where sludge had escaped from the insured's sewage plant, contaminating a river.

8.4. Employer's liability insurance

In Great Britain, employer's liability insurance is mandated by the Employers' Liability (Compulsory Insurance) Act 1969. The insurance covers liability of the employer for illness, injury or death suffered by their employees.

Compensation for work-related diseases (e.g. dermatitis and mesothelioma) can be paid to former employees. However, work-related motor accidents are covered by compulsory motor insurance.

The statutory minimum amount of cover is £5 million for any one occurrence. Recovery is 'fault-based', as the employer has to be legally liable for the illness or injury (e.g. when the employer has been negligent).

Alternatively, other jurisdictions provide 'no fault-based' recovery, for example, government-funded social security in New Zealand, France, Italy, Spain and Germany, and compensation funded by employer-purchased insurance policies in the USA and Australia.

8.5. Professional indemnity insurance

Professional indemnity (PI) insurance covers awards of damages, costs or settlements resulting from a claim made during the policy period for any act, error or omission arising out of the conduct of the business. The insured does not need to have been negligent to have a successful claim made against them.

PI insurance is described as a 'claims-made' basis as cover, for an event that may have occurred many years before, is for matters notified to the insurer during the policy period. The policy will be subject to a limit of indemnity and should be on an each and every claim basis.

To ensure cover, the insured's business should be comprehensively detailed on their proposal form. Many professional bodies require their members to maintain PI insurance.

8.5.1 Professional negligence claim against a surveyor

Box 8.3 *Large v Hart & Anor* [2021] EWCA Civ 24

The Court of Appeal reviewed the quantification of loss in a surveyor's professional negligence claim. The surveyor had been engaged to survey a property, which had been rebuilt and extended, and produce a RICS HomeBuyer report.

The Technology and Construction Court (TCC) judge, at first instance, found that the surveyor was negligent for failing to identify and report the extensive damp at the property, and he should have recommended that a professional Consultancy Certificate be obtained prior to purchase. The TCC awarded damages for the difference between the value of the property as stated in the RICS HomeBuyer report and its value with all the defects that existed.

The surveyor appealed on the correct measure of assessing loss. The Court of Appeal ruled that the TCC's calculation of loss was appropriate and correct.

However, the case in Box 8.3 confirms that the traditional measure of assessing loss in surveyors' claims – the diminution of value test (a comparison between the value of the property in the condition that was reported and the value it should have been reported to be in) – remains good law in the majority of cases.

8.5.2 Distinction between reasonable skill and care and fitness-for-purpose design

The common law duty of care required of a construction professional, and also implied under the Supply of Goods and Services Act 1982, is to take reasonable skill and care.

Box 8.4 *Bolam v Friern Hospital Management Committee* [1957] 1 WLR 582

The judgement in this case (at p. 586) sets out the 'Bolam test of reasonable skill and care' as

> the standard of the ordinary skilled man exercising and professing to have that special skill.

This case was related to an incident at the hospital when the patient, Bolam, received electroconvulsive therapy which caused him to suffer serious bone fractures. Bolam argued that his doctor had been negligent. However, the claim failed, as it was decided that the doctor had followed the medical protocol at the time, and patients were also not routinely advised of all the small risks that could occur as a result of the procedure.

In contrast, a fitness-for-purpose standard imposes a higher duty, as it is an absolute obligation to achieve a stipulated specification. This duty comes from the Sale of Goods Act 1979.

A design-and-build contractor, in the absence of an expressed contractual rebuttal, must ensure that the works completed are fit for their intended purpose.

Box 8.5 *Independent Broadcasting Authority v EMI Electronics Ltd and BICC Construction Limited* [1979] 11 BLR 29

The Court of Appeal judges stated (at p. 34)

> We see no good reason … for not importing an obligation as to reasonable fitness for purpose into these contracts.

In this case, the claimant, the Independent Broadcasting Authority (IBA), had engaged the first defendant (EMI) to design and build three cylindrical aerial masts. The second defendant (BICC) was the nominated subcontractor and had carried out the design for the masts.

In 1969, the first of the three masts broke and collapsed. IBA commenced proceedings against EMI for breach of contract and negligence, and also against BICC for negligence and breach of warranty and negligent misstatement. It had been argued and accepted that the mast was 'both at and beyond the frontier of professional knowledge at that time'.

However, *Keating on Construction Contracts* (Furst and Ramsay, 2016, p. I-040) advises that 'the nature of the design obligation will depend on the terms of the contract and it is desirable to define the design obligation in an express term'.

A contractor may take on a fitness-for-purpose liability, which it cannot pass on to its design team, who will usually accept only a reasonable skill and care standard of responsibility (Lupton, 2013). Furthermore, *Hudson's Building and Engineering Contracts* (Atkin Chambers, 2015) explains that the courts have consistently rejected contractors' arguments that their scope of liability for design should be one of due or professional care only, on the basis that it is unfair to impose a higher duty than that of a professional designer.

Professional indemnity policies significantly influence the design standard because they will cover professional negligence (i.e. a failure to exercise reasonable skill and care) but normally exclude protection for fitness-for-purpose obligations (Lupton, 2013). Therefore, a designer or contractor may be left uninsured against a contractual claim for breach of a fitness-for-purpose obligation.

Charlson (2019) covers contractors' standard of design responsibility by analysing current standard forms of contract and reviewing recent relevant case law.

8.6. Structural warranties

A structural warranty is an insurance policy designed to protect against defects in new buildings, for a defined period known as the 'policy term'. While a structural warranty is purchased by the contractor or developer before construction begins, it protects the owner from structural damage that may occur during the period of insurance following its completion. Warranties can also be transferable in the period of insurance to benefit subsequent owners.

The cover provided by a structural warranty can include

- structural insurance – the main part of the warranty, which covers against structural issues after the defects period expires

- deposit protection – should the developer become insolvent during construction, this protects the purchaser's deposit
- defects insurance – may include cover for non-structural issues
- contaminated land – protects against the cost of removing contamination
- building control cover – cover if the property was not built in compliance with the Building Regulations 2010.

Most mortgage lenders and banks require a structural warranty to be in place before they will loan money secured by a mortgage on a property. Main factors that mortgage lenders look for in structural warranties include

- the length of the cover period (the industry standard is currently 10 years)
- the process to ensure the build quality and compliance with the Building Regulations 2010, such as the inspection regime
- the financial limits of the cover
- the consistency of cover of the period (often an initial period is covered by the contractor, after which it reverts to the warranty provider alone – lenders would want to ensure a consistent level of protection)
- what elements of cover are included (e.g. contractor's insolvency)
- information which provides evidence of financial stability
- confirmation that the warranty provider (and broker, if relevant) has evidence of Financial Conduct Authority authorisation
- any evidence of claims experience.

8.7. Non-negligent liability

Various Joint Contracts Tribunal (JCT) contract conditions mandate non-negligent liability insurance to protect the client against liability for damage to adjacent or surrounding property.

Box 8.5 *Gold v Patman & Fotheringham Ltd* [1958] EWCA Civ J0522-3

Such a requirement arose following this case, which established the principle that the client is liable in tort for damage to third-party property, where this is not a consequence of the negligence of the contractor.

Gold was the Employer and Patman & Fotheringham was the Contractor engaged on a standard RIBA form. Damage was caused to an adjoining neighbour's property due to piling.

The neighbour brought action against the Employer. The Employer sought to recover against the Contractor. Under the contract conditions, the Contractor was only liable for damage if negligence was established.

The court decided that the damage was not attributable to the Contractor's negligence. The Employer was held liable in nuisance for removing support to the neighbour's land. Therefore, the Employer was liable for the costs of the damage, and with no insurance protection.

The key outcomes as a result of *Gold v Patman & Fotheringham Ltd* were

- public liability insurance will not pay out if negligence cannot be proved
- injured party/parties can sue the developer or employer that brought the contractor onto site.

As a consequence, the 'non-negligent damage' clause was invented.

Non-negligent liability insurance is usually purchased by the contractor on behalf of the client, because the contractor's annual public liability policy is better suited to providing such an indemnity. The cover can be arranged in the joint names of the contractor and client.

Specific risks identified in JCT contracts include

- collapse
- subsidence
- heave
- vibration
- weakening or removal of support
- lowering of groundwater.

Non-negligence insurance is not always required (e.g. where construction is on a greenfield site). However, it should be considered where piling, underpinning or basement works are being undertaken in suburban or built-up areas.

It should be noted that non-negligence liability insurance has several key exclusions, which are

- loss or damage arising from negligence (otherwise insured by the contractor's public liability policy)
- damage which is considered to be inevitable as a result of the carrying out of the works (not an insurable risk)
- loss or damage attributable to the designing of the works (usually insured under a professional indemnity policy – carried by the contractor or designer)
- loss or damage otherwise insured under buildings and/or contract works insurance policies.

8.8. Directors' and officers' liability insurance

Directors' and officers' (D&O) liability insurance cover is for awards of damages, costs or settlements for which a director or officer is liable, resulting for a claim made against them during the policy period for any act, error or omission in their capacity as a director or officer of the company. Fines and penalties are typically excluded.

D&O cover is taken out in the company name and is a 'claims-made' cover (i.e. only claims notified during the policy period are covered).

8.9. Environmental impairment liability insurance

Environmental insurance can be defined as protection against an insured's exposure to the liabilities arising from the ownership and/or development of brownfield land and any consequential loss and damages. Cover includes exposure to

- third-party claims, for example, property damage or personal injury
- regulatory action by an authority that requires investigation, clean-up of contaminated ground/water or the restoration of environmental damage.

Traditional insurance policies generally exclude cover for environmental pollution. The insurance market is wary of covering environmental risks as environmental pollution tends to be gradual, taking many years to manifest, which makes the risk difficult to quantify.

The original cause of pollution may be long forgotten and the perpetrator no longer in business, which makes the recovery of any insured costs from third parties more difficult. Furthermore, the insurance industry has already paid out on some enormous environment-related claims, such as asbestosis claims in the USA.

8.9.1 Availability of environmental insurance

Whether it is possible to obtain environmental insurance for contamination risks depends on a number of factors, including

- former and current land use and operations
- the likelihood of remediation being required
- other risks, such as the probability of pollution or further contamination occurring.

Insurers therefore need to evaluate the risk before underwriting environmental insurance. Environmental reporting can be undertaken in three stages.

1. Historical data are collected and interrogated in 'desk-top' reporting. Sources can include previous uses and geological maps, local authority information (e.g. proximity to landfill sites), previous environmental incidents and planning history.
2. A site investigation may include taking soil and water samples and digging trial pits or boreholes.
3. A final stage can be remediation and validation, such as post-completion monitoring of ground gas or water.

8.9.2 Exposure to environmental claims

A contractor, engaged on traditional lump-sum contracts, is vulnerable to claims arising out of environmental non-compliance (e.g. restoration costs following pollution of a watercourse). Furthermore, a contractor may also be expected to accept liability for pre-existing contamination on brownfield sites.

Environmental exposure for owners of brownfield and contaminated land can include historical pre-existing or new gradual pollution. Separate policies can be purchased to cover pre-existing pollution (historical contamination) and new pollution (future contamination). However, insurance cover is not available for fines imposed for criminal conduct (e.g. fines consequential to a breach of environment statutes or regulations).

8.9.3 Environmental insurance options

Most policies only provide aggregate cover for polluting events, which means the cover could be exhausted by one large claim, leaving the insured with no cover for the remainder of the policy year. Options include annual or long-term policies of up to 25 years.

Some insurers offer a policy option for a particular project, which consists of a fixed-price premium that is freely transferable between specified parties for a defined period. This means that insurance can be transferred on sale of all or part of the land. However, environmental warranties and indemnities remain the principal risk allocation tools, rather than environmental insurance, to support large commercial transactions.

Environmental impairment liability insurance tends to be written on a 'claims-made' basis, which avoids the problem of defining when a pollution incident has occurred. The main types of specialist environmental cover currently available include

- pollution liability insurance, which can protect against losses associated with known and unknown pollution, including historic contamination
- remediation cost cap insurance, which can be obtained to cover cost overruns in carrying out remediation of contamination
- contractors' pollution liability insurance, which can be taken out to cover the risk of pollution being caused during construction or remediation projects.

8.9.4 Reporting of polluting incidents

Generally, there are no reporting requirements under environmental legislation in England and Wales. As a consequence, in circumstances where a polluting incident occurs, the polluter is usually not obliged to report the incident to the regulatory authorities.

The fact there are no reporting requirements does not mean that a polluter will avoid detection, liability for clean-up costs, or even criminal liability provided for under environmental legislation. The lack of a requirement to report does not prevent a polluter from reporting an incident voluntarily.

Where the incident is a serious one, with an imminent threat of serious harm being caused to the environment, it may be wise to be proactive by reporting the incident to the regulatory authorities. The polluter will then have the opportunity to liaise and cooperate with the regulatory authority in drawing up an action plan to clean up the site to a standard required under environmental legislation.

Proactivity is not a guarantee of avoiding the requisite penalties provided for under environmental legislation. However, it may be sufficiently persuasive to convince regulatory authorities to issue a lesser penalty. Any mitigating factors, such as prompt admission of liability or an undertaking to remedy the situation, may assist the regulatory authorities in how they should exercise their discretion in relation to a specific offence.

In the event that the incident is of a minor nature and can be managed without advice from the regulatory authorities, it may be thought that there is no need to report the incident.

8.10. The Chancery Lane Project – climate clauses

The Chancery Lane Project (TCLP, 2022a) is a collaborative initiative of international legal and industry professionals whose vision is a world where every contract enables solutions to climate change.

The project creates new, practical contractual clauses ready to incorporate into law firm precedents and commercial agreements to deliver climate solutions. The TCLP's climate clauses (TCLP, 2022b) include many with an insurance focus, for example

- repair or refurbishment in insurance claims
- climate-related knowledge sharing between insurer and insured
- incentives for insured parties to disclose and meet greenhouse gas emissions targets
- general condition to commercial insurance policies: climate change risk assessment
- green dispute reporting in after-the-event insurance policies
- insurance: disclosure and mitigation of pending climate change litigation
- exclusions from insurance coverage for climate harms.

TCLP then works with lawyers to ensure effective and impactful implementation of the clauses across industries, practice areas and jurisdictions.

However, there is a risk that some climate change clauses could be interpreted as imposing fitness-for-purpose responsibilities. As explained in Section 8.5.2 above, a fitness-for-purpose obligation is usually excluded from professional indemnity insurance cover.

References

Statutes

Employers' Liability (Compulsory Insurance) Act 1969
Sale of Goods Act 1979
Supply of Goods and Services Act 1982
Water Resources Act 1991

Regulations

The Building Regulations 2010

Case law

Bartoline Limited v Royal & Sun Alliance Insurance plc [2006] EWHC 3598 (QB)
Bolam v Friern Hospital Management Committee [1957] 1 WLR 582
Gold v Patman & Fotheringham Ltd [1958] EWCA Civ J0522-3

Independent Broadcasting Authority v EMI Electronics Ltd and BICC Construction Limited [1979] 11 BLR 29

Large v Hart & Anor [2021] EWCA Civ 24

Yorkshire Water Services Ltd v Sun Alliance & London Insurance plc and Others [1997] 2 Lloyd's Rep. 21

Journals and books

Atkin Chambers (2015) *Hudson's Building and Engineering Contracts*, 13th edn. Sweet & Maxwell, London, UK.

Charlson J (2019) Briefing: Interpreting contractors' mandated standard of design. *Proceedings of the Institution of Civil Engineers – Management, Procurement and Law* **172(4)**: 142–145. 10.1680/jmapl.19.00007.

Furst S and Ramsay V (2016) *Keating on Construction Contracts*, 10th edn. Sweet & Maxwell, London, UK.

Lupton S (2013) *Cornes and Lupton's Design Liability in the Construction Industry*, 5th edn. Wiley-Blackwell, Oxford, UK.

Websites

TCLP (The Chancery Lane Project) (2022a) About The Chancery Lane Project. https://chancerylaneproject.org/about (accessed 09/03/2023).

TCLP (2022b) Climate clauses. https://chancerylaneproject.org/climate-clauses (accessed 09/03/2023).

Further reading

De Silva C and Charlson J (eds) (2021) *Galbraith's Construction and Land Management Law for Students*, 7th edn. Routledge, London, UK.

RICS (2021) *Risk, Liability and Insurance*, 1st edn. RICS, London, UK.

Baker F and Charlson J
ISBN 978-0-7277-6645-8
https://doi.org/10.1680/elsc.66458.195

Chapter 9
Proactive project planning

Jennifer Charlson

This chapter incorporates indicative checklists of tasks to be undertaken at pre-contract, on-site and post-contract stages. The relevant previous chapters are cross-referenced. However, please note that these checklists are not exhaustive, as each individual project will have specific concerns.

For ease of reference, the previous chapters are listed here

1. The UK environmental legal framework, regulators and advisers
2. Planning and environmental permits
3. Environmental impact and habitat assessments
4. Contaminated and brownfield land
5. Waste management
6. Water pollution
7. Site hazards and nuisances
8. Insurance

Pre-contract checklist

Task	Chapter
• Does the proposed development require planning permission?	2
• Alternatively, do permitted development rights apply?	2
• Is a change of use application necessary?	2
• Is demolition permission required?	2
• Is a Community Infrastructure Levy payable?	2
• Are any environmental permits required?	2
• Is a design and access statement required?	3
• Is it necessary to carry out an environmental impact assessment?	3
• Do the Habitat and Species Regulations apply to the development proposal?	3
• Consider biodiversity net gain?	3
• Is a noise management plan required?	3
• Is a flood risk assessment necessary?	3
• Does the site present contaminated land risks?	4
• Is the site brownfield land?	4
• Is a waste management strategy in place?	5
• Has hazardous waste been identified?	5
• Has the cost of waste disposal and/or landfill tax been estimated?	5
• Are water or groundwater discharge permits required?	6
• Will there be water discharge to sewers?	6
• Are new connections to public sewers planned?	6
• Is drainage tax payable?	6
• Is a water abstraction licence needed?	6
• Is nutrient pollution a concern?	6
• Is there adequate insurance cover?	8
• Has environmental insurance been considered?	8

On-site checklist

Task	Chapter
• Check for planning permission compliance	2
• Is the site secure from trespassers, in particular children?	7
• Ensure compliance with the Control of Substances Hazardous to Health Regulations 2002	7
• Ensure compliance with hazardous waste regulations	7
• Has consideration been given to minimising noise, smell and dust emissions from site?	7
• Ensure compliance with the Control of Pollution Act 1974 (Noise)	7
• Check for potential public and private nuisances	7
• Could foreseeable damage be caused by escapes occasioned by the non-natural use of land?	7
• Consider potential liability in negligence	7
• Have potential hazards which may be posed by the site been assessed?	7
• Has the proximity of watercourses been checked for both surface and ground water?	6
• Are there adequate protection measures in place to prevent water pollution, in particular by nutrients (e.g. phosphate and nitrate)?	6
• Are there adequate procedures in place for the management of waste?	5

Post-contract checklist

Task	Chapter
• Check that contaminated or brownfield land has been remediated	4
• Is there a contractual obligation to maintain adequate insurance cover in respect of work undertaken or advice given?	8

Baker F and Charlson J
ISBN 978-0-7277-6645-8
https://doi.org/10.1680/elsc.66458.199

Table of legal sources

Statutes

Regulations

Directives

Directions

Orders

Case law

Baker F and Charlson J
ISBN 978-0-7277-6645-8
https://doi.org/10.1680/elsc.66458.205

Index